T0184146

# THE ROLE OF MATHEMATICS DISCOURSE IN PRODUCING LEADERS OF DISCOURSE

A Volume in
The Montana Mathematics Enthusiast Monograph Series in Mathematics Education

*Series Editor:*
Bharath Sriraman
The University of Montana

# The Montana Mathematics Enthusiast
# Monograph Series in Mathematics Education

Bharath Sriraman
*Series Editor*

*Interdisciplinarity, Creativity, and Learning: Mathematics with Literature,*
*Paradoxes, History, Technology, and Modeling* (2009)
Edited by Bharath Sriraman, Viktor Freiman, and Nicole Lirette-Pitre

*Critical Issues in Mathematics Education* (2009)
Edited by Paul Ernest, Brian Greer, and Bharath Sriraman

*Relatively and Philosophically E³rnest*
*Festschrift in honor of Paul Ernest's 65th Birthday* (2009)
Edited by Bharath Sriraman and Simon Goodchild

*Creativity, Giftedness, and Talent Development in Mathematics* (2008)
Edited by Bharath Sriraman

*Interdisciplinary Educational Research in Mathematics and Its Connections*
*to the Arts and Sciences* (2008)
Edited by Bharath Sriraman, Claus Michelsen,
Astrid Beckmann, and Viktor Freiman

*Mathematics Education and the Legacy of Zoltan Paul Dienes* (2008)
Edited by Bharath Sriraman

*Beliefs and Mathematics:*
*Festschrift in Honor of Guenter Toerner's 60th Birthday* (2007)
Edited by Bharath Sriraman

*International Perspectives on Social Justice in Mathematics Education* (2007)
Edited by Bharath Sriraman

# THE ROLE OF MATHEMATICS DISCOURSE IN PRODUCING LEADERS OF DISCOURSE

**Libby Knott**

*Washington State University*

**INFORMATION AGE PUBLISHING, INC.**
Charlotte, NC • www.infoagepub.com

**Library of Congress Cataloging-in-Publication Data**

Knott, Libby.
  The role of mathematics discourse in producing leaders of discourse /
Libby Knott.
      p. cm. – (The Montana mathematics enthusiast: monograph series in
mathematics education)
  Includes bibliographical references.
  ISBN 978-1-60752-282-9 (pbk.) – ISBN 978-1-60752-283-6 (hardcover) –
ISBN 978-1-60752-284-3 (e-book)
1.  Mathematics–Study and teaching.  2.  Mathematics teachers–In-service
training.  3.  Curriculum planning.  4.  Discourse analysis.  I. Title.
  QA13.K65 2009
  510.71–dc22
                              2009035542

Printed in the United States of America

# CONTENTS

# PREFACE TO THE ROLE OF MATHEMATICS DISCOURSE IN PRODUCING LEADERS OF DISCOURSE

I am pleased to present the ninth monograph in the Montana Mathematics Enthusiast Monograph Series in Mathematics Education,[1] guest edited by Libby Knott of Washington State University. Dr. Knott has been involved in research on discourse analysis in mathematics education for over a decade and this monograph represents a compilation of her efforts in this domain of research.

This volume comprises a collection of related chapters concerning the implementation of discourse in the mathematics classroom, its forms and functions. Chapters 1, 3, 4, 5, 8, 10, and 11 feature work that emanated, at least in part, from work done in conjunction with the Oregon Mathematics Leadership Institute. This was a National Science Foundation-funded Mathematics and Science Partnership Teacher Institute grant that was directed at K–12 teachers in Oregon. The grant provided a residential sum-

[1] The Oregon Mathematics Leadership Institute Partnership Project is funded by the National Science Foundation's Math Science Partnership Program (NSF-MSP award #0412553) and through the Oregon Department of Education's MSP program.

mer institute for three summers in mathematics content and collegial leadership for participants.

One of the unique characteristics of the program were that K–12 teachers learned mathematics content—geometry, discrete math, data and chance, number and operations, measurement, and change, algebra—in homogeneous classrooms. The glue that held the many different facets of the institute together was its emphasis on mathematical discourse, particularly on generalization and justification. These two features were emphasized, modeled, and researched in all six of the mathematics content classes. They figured largely in the discussions in the leadership classes, with the focus being on how these teachers would take back what they had learned to their colleagues in their schools and districts and help them to emphasize generalization and justification in their discourse and in that of their students.

Chapter 1 provides a description of one of the institute content courses, and concentrates on providing a comparison of implementing this type of course during the summer institute, with a similar implementation during a college semester. During the summer the course was team taught, while during the year the course was taught with a single instructor. The authors conclude that the presence of a second instructor impacted the cognitive level of the discourse. Chapter 2 provides details of how one might provide a discourse-rich calculus class, particularly how one might successfully steer the student discourse towards reflective abstraction. Chapter 3 provides comparisons of different instruction in sections of a pre-calculus class, and argues that there are three important factors that can influence student participation and achievement, not only the classroom discourse but also the teacher beliefs and the classroom environment. Chapter 4 gives a first-hand look into the third-grade classroom of a teacher who participated in OMLI and was eager to put into practice what she had learned. This gives a rich description of what can be accomplished. In chapter 5, we are given an up-close look at one of the discourse-rich classrooms that were offered during the institute. We can see clearly what the norms were in such a classroom, what was expected of the teacher participants in terms of mathematical discourse that emphasized generalization and justification. In chapter 6, we are provided with an example of how one might work with pre-service teachers on orchestrating discourse in the classroom and some of the difficulties one might encounter. Chapter 7 again looks at orchestrating discourse, this time in a whole-class setting. The setting is a middle-school classroom and we are provided with some concrete examples of how a teacher implements whole-class discourse while pressing students to investigate mathematical ideas. Chapter 8 looks again at implementation of discourse in the classroom but focuses on the disconnect that sometimes occurs between intended curriculum and enacted curriculum, and shows how difficult it can be for teachers to enact their lesson as written, and maintain a high cognitive level of discourse. Chapter 10 again provides us with a close-up view of the development and enactment of a discourse-rich geometry class during OMLI summer institute. In particular, a compare and contrast approach was taken to the curriculum development. The final chapter looks at the next layer of complexity—that of the work of professional development leaders and describes some of the obstacles they face when attempting to implement work focusing on mathematical justification with teachers. The authors amply illustrate the necessity of mathematical knowledge for teaching and the establishment of socio-mathematical norms in promoting successful professional development work with teachers.

Bharath Sriraman
*Series Editor*
*The Montana Mathematics Enthusiast*
*Monograph Series in Mathematics Education*
Missoula, Montana
March 12, 2009.

CHAPTER 1

# STUDENT MATHEMATICAL DISCOURSE AND TEAM TEACHING

**Martha VanCleave and Julie Fredericks**

It has been proposed that student achievement in mathematics can be significantly improved by increasing the quality and quantity of meaningful mathematics discourse in the classroom. This paper explores student mathematical discourse in both undergraduate mathematics classrooms and summer-institute classes for inservice teachers. Utilizing a discourse observation protocol, comparisons between the quality of discourse in these two settings led to a further investigation of the impact of team teaching on student mathematical discourse. The actions of the second instructor in the team teaching setting classified as pressing for clarification, extension, side-trip, and highlighting, appear to increase the proportion of high level student mathematical discourse.

**KEYWORDS:** student mathematical discourse; team teaching; mathematics professional development

## INTRODUCTION

This paper compares the level of student mathematical discourse in two different classroom settings: a summer institute for K–12 teachers and under-

*The Role of Mathematics Discourse in Producing Leaders of Discourse*, pages 1–15

graduate mathematics courses.[1] The same instructors led the classes in both settings. In preparation for the summer institute, the instructors attended workshops focused on the importance of and methods for eliciting student mathematical discourse. After participating in the summer institute, the instructors returned to teaching their typical undergraduate courses. The initial research question focuses on whether the classrooms in the two settings exhibit similar levels of mathematical discourse.

## BACKGROUND ON MATHEMATICAL DISCOURSE

The focus on discourse in the teaching of mathematics, which has been the subject of considerable research in K–12 teaching, is extending to undergraduate mathematics education. The research on discourse in K–12 mathematics teaching has identified the characteristics of quality discourse and its impact on student learning (NCTM, 1989, 1991, 2000, 2003). Research is now examining undergraduate mathematics teaching in order to determine the relevance of the K–12 findings in this setting.

While it is important for students to be communicating, not all types of student mathematical discourse have the same impact on development of students' conceptual understanding. When student discourse is oriented toward what one *does* rather than what one *thinks*, students may continue to believe that the reason they are given mathematical problems is to find answers that are numbers or calculations, or that working problems means searching for operations to perform. They also may not be able to assess their own understanding of mathematical concepts (Clement, 1997). A classroom environment where students are expected to go beyond explaining how they solve a problem to sharing their mathematical understanding can be established by the use of socio-mathematical norms. Yackel (2001) cites the importance of socio-mathematical norms of explanation and justification. These norms include "that students explain and justify their thinking, that they listen to and attempt to make sense of the explanation of others, and that explanations describe actions on objects that are experientially real for them" (Yackel, 2001, p. 1–17). Further, Yackel emphasizes that the instructor and the students develop the norms interactively. The instructor sets the expectations and influences the ways in which students (especially shy and reticent students) participate in the discussion. The students contribute to the norms by increasingly acting in accordance with these expectations.

---

[1]The Oregon Mathematics Leadership Institute Partnership Project is funded by the National Science Foundation's Math Science Partnership Program (NSF-MSP award #0412553) and through the Oregon Department of Education's MSP program.

## COMPARISON OF SUMMER INSTITUTE & REGULAR CLASSROOMS

*The Settings*

The motivation for this study grew out of our experience working in a National Science Foundation Math Science Partnership (NSF-MSP) teacher institute project. The project represented a partnership between two state universities, 10 public school districts, and a non-profit professional development organization. The project staff included higher-education faculty from mathematics and education departments at public and private colleges, universities and community colleges; K–12 master teachers; and experienced professional development staff. The teachers participating in the project spanned all grade levels, from kindergarten through high school. One of the project's central activities was a series of three-week institutes over three consecutive summers. Each participating teacher attended two mathematics content courses each summer (for a total of six content courses throughout the project.) Each of these content courses was team taught (by four instructors) with a class size of approximately 30 and usually split into two sections of about 15 participants.

Goals of the project were to:

- Increase mathematics achievement of all students in participating schools;
- Close achievement gaps for underrepresented groups of students; and
- Increase enrollment and success in challenging mathematics course work that supports state and national standards through coherent, evidence-based programs.

In order to achieve these goals, one focus of the project was to build the strong content knowledge

> necessary to enable teachers to transform their classes into mathematical learning communities where students engage in high level discourse around important mathematical ideas. Teachers' content knowledge must be built in ways that connect important ideas clearly to school mathematics and using approaches that model effective instruction (Dick, 2004, p. 1).

Specifically, the project's logic model proposed that student achievement in mathematics could be significantly improved by increasing the quantity and quality of meaningful mathematics discourse in the mathematics classes in the schools of the teacher participants. Hence, attention to the teacher participants' mathematical discourse was also stressed in the summer institute mathematics content classes.

*Pilot Investigation: Impact of Summer Institute Experience on Higher Education Faculty*

A key evaluation question for the project involved documenting the quantity and quality of student mathematical discourse in a sample of the teacher participants' classrooms to determine the project's impact on those classrooms. We hypothesized that the project might have a similar impact on the regular undergraduate classrooms of the higher-education faculty teaching the institute mathematics content courses in which they paid particular attention to the participant mathematical discourse. With that in mind, we undertook a pilot case study of the student mathematical discourse in the regular classes of three members of the grant project instructional staff. These classes took place in a private university, a private college, and a public community college. The classes represented regular teaching assignments of the faculty. Classes included developmental mathematics, mathematics for elementary teachers, calculus, discrete math, and probability and mathematical statistics. All of the classes were small, fewer than 30 students, and each was taught by a single instructor.

The primary subjects for this study were three higher-education grant project staff teaching in the same content course during the summer institute, one of whom is an author of this study. All subjects hold doctorates in mathematics. One of the subjects is a full time faculty member at a community college and has been teaching at the college level for over five years. The other two subjects are professors at private colleges, one a full professor with over 30 years college teaching experience and the other an assistant professor with over five years experience. All three subjects collaborated with a master teacher on the development of the institute content course and team-taught the course in pairs.

*Methodology*

For our case studies, observations of all three subjects were conducted during one summer institute of the grant projects and in their regular undergraduate classes the following fall term. During the summer institute, observations of two sessions led by each instructor were conducted by student research associates. During the fall term, two or three observations of each class being taught by each instructor were conducted by one of the authors. Each observation was videotaped and data on the quality and quantity of discourse was recorded using the discourse observation protocol developed by the grant project (Weaver & Dick, 2006). All observers were trained in the use of the discourse observation protocol.

The discourse observation protocol was developed specifically to record and measure the quantity and quality of student mathematical discourse in

## Table 1: Discourse Taxonomy Used Classifying Levels of Discourse[1]

| Level | Type of Discourse | Description |
|-------|-------------------|-------------|
| 1 | Answering, stating, or sharing | A student gives a short right or wrong answer to a direct question, or a student makes a simple statement or shares his or her results in a way that does not involve an explanation of how or why. |
| 2 | Explaining | A student explains a mathematical idea or procedure by describing how or what he or she did but does not explain why. |
| 3 | Questioning or challenging | A student asks a question to clarify his or her understanding of a mathematical idea or procedure, or a student makes a statement or asks a question in a way that challenges the validity of an idea or procedure. |
| 4 | Relating, predicting, or conjecturing | A student makes a statement indicating that he or she has made a connection or sees a relationship to some prior knowledge or experience, or a student makes a prediction or a conjecture based on his or her understanding of the mathematics behind the problem. |
| 5 | Justifying or generalizing | A student provides a justification for the validity of a mathematical idea or procedure, or makes a statement that is evidence of a shift from a specific example to the general case. |

[1]From Oregon Mathematics Leadership Institute Spring 2007 Evaluation Report by D. Weaver, 2007, Portland, OR, RMC Research Corporation, p. 26.

the classroom. The aspect of the observation protocol analyzed in this study involves the discourse taxonomy (see Table 1) that classifies each incident of student discourse by types that are grouped in levels. The first level of discourse including answering, stating, or sharing is the lowest cognitive demand level. Discourse in Level Five including justification and generalizing is the highest cognitive demand level. The observation protocol also records size of group. Only data from whole class episodes is presented in this paper. The mode of discourse and tools used are also recorded but are not examined in this study.

### Data Analysis

The analysis began with an examination of the student mathematical discourse data for each instructor. The summer institute data for each instructor were drawn from the two observed sessions led by that instructor. Similarly, the regular classroom data for each instructor were drawn from all observed classes of that instructor. Each instructor was teaching at least two different math courses during the observation period. Level of discourse was analyzed by computing the proportion of discourse at each level for each instructor. Comparisons were made between the proportions of

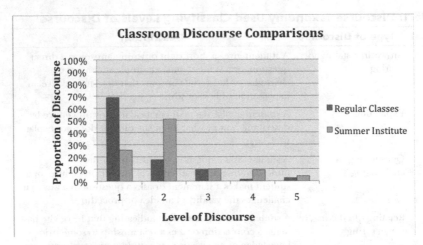

**FIGURE 1. Level of discourse comparisons between regular and summer institute classes.**

discourse at each level for all three instructors. The trends in the distributions of proportions of discourse were similar for all three instructors. To facilitate the analysis of the comparison between the two settings, the data from all instructors were combined for each setting and the same trends seen for individual instructors are still present.

Comparisons between proportions of levels of discourse in the two settings are justified by similar rates of discourse (1.72 instances per minute in the regular classes and 1.21 instances per minute in the summer institute). In regular classes, Level One discourse is the mode while in the summer institute course, Level Two discourse is the mode (see Figure 1). Further, in regular classes the proportion of discourse decreases as the cognitive level of discourse increases. Clearly, since Level Two is the mode for the summer institute course the decreasing trend is not present. Moreover, the proportion of discourse at the higher cognitive levels (Four and Five) tends to be greater in the summer institute as compared to the regular classes of all instructors.

## Discussion and a New Hypothesis on the Impact of Team Teaching on Student Discourse

The contrast between what was seen in the summer institute classrooms and the regular classrooms of the project instructors reveal differences in the proportions of discourse in these two settings. Further, the data collected by the NSF-MSP project on student mathematical discourse in the teacher participants' classrooms, while showing change from year to year,

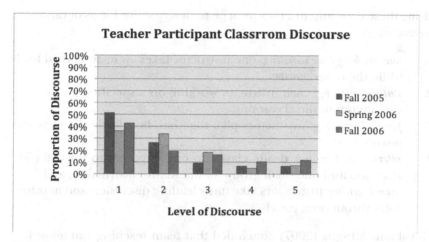

**FIGURE 2. Level of discourse in teacher participant classrooms.**

still showed a decreasing trend from Level One through Level Five similar to the regular classrooms of the instructors (see Figure 2).

The observation that both the regular classrooms of the higher education faculty and the classrooms of the teacher participants displayed similar distributions of discourse that differed substantially from that of the summer institute classrooms led to further examination of the summer institute classes.

A team of two instructors facilitated each class in the summer institute. For each lesson, one instructor was the lead and was primarily responsible for facilitating the activities and discussion. The second instructor acted in a supporting role assisting with the monitoring of small groups, time management, selection and sequencing of presentations and facilitation of discussion. Since team teaching was such an integral part of the summer institute classrooms and not present in the regular classrooms of either the instructors or the teacher participants, it was conjectured that the structure of the summer institute, particularly the employment of team teaching, may have had an impact on the level of student mathematical discourse present in these classrooms.

## BACKGROUND ON TEAM TEACHING

In their research on team teaching Cook and Friend (1996) identified four key components of co-teaching:

1. two educators;
2. delivery of meaningful instruction;
3. diverse groups of students; and
4. common settings.

Utilizing these components, they went on to describe five forms of variation in co-teaching:

1.  *one teaching/one assisting*: one instructor takes an instructional lead while the other assists;
2.  *station teaching*: each instructor working on a specified part of the curriculum in the classroom;
3.  *parallel teaching*: instructors plan together, but divide the class for instruction;
4.  *alternative teaching*: divide class into one large group for main instruction and one small group for alternative instruction; and
5.  *team teaching*: instructors take turns leading discussion and in other roles throughout the class.

Grassl and Mingus (2007) concluded that team teaching can allow for dynamic interaction between the instructors and between instructors and students, allowing students to experience different viewpoints of the instructors. Some advantages of team teaching cited in their study include: students hearing alternative ways of explaining the same concept; the availability of immediate feedback on how the class is progressing; the assisting instructor asking leading questions to clarify student thinking, make extensions, and/or connections; and the assisting instructor highlighting opportunities for student speaking. In addition, Grassl and Mingus found evidence that team teaching supported efforts to sustain reform teaching beyond the team teaching setting. Both instructors involved in their study taught the subsequently taught the course independently and noted that one of them taught the course with the "same spirit, organization and results" while the other instructor has "changed the nature of her exams to included more challenging problems, with higher expectations" (Grassl & Mingus, 2007, p. 596).

The summer institute course embodied the four components of team teaching stated above and primarily used the one teaching/one assisting form described. Within this framework, we formulated a new research question: Does team teaching foster the increased level of student mathematical discourse observed in the summer institute classroom by affording an advantage similar to those suggested by Grassl and Mingus?

## TEAM TEACHING IN THE SUMMER INSTITUTE CLASSROOMS

*Re-Examination of the Data*

The videotapes of the summer institute classes were reviewed with a focus on the actions of the second instructor. This review identified four categories of second-instructor actions: monitoring small groups, time management, selecting and sequencing, and interjections into whole class discussion. While all of these second-instructor actions added to the effectiveness of the instructional team, the direct effect on student discourse can only be isolated in the interjections to whole class discussion. The second instructor actions during the monitoring of small groups likely affected the level of discourse within the small groups, but cannot be shown to have affected discourse in whole group discussions. The skills developed by both instructors through working together on the selecting and sequencing of presentations appear to be readily transferable to the solo instructor classroom. Therefore, the episodes in which the second instructor made interjections into whole class discussions were further examined. These episodes were transcribed to facilitate detailed analysis of the student mathematical discourse and the role of the second instructor.

To determine whether the actions of the second instructor during the whole class discussion raised the level of student mathematical discourse, the student discourse following interjections by the second instructor during the whole class discussion were recoded. This recoding was necessary because the discourse in these episodes could not be isolated in the original observation discourse protocol records. Analysis of this recoding of student discourse revealed an entirely different distribution of levels of discourse following an interjection by the second instructor than was found in general whole class discussion during the summer institute class. The primary difference in these distributions was higher proportions of student discourse in Levels Three, Four and Five than were seen in other situations (see Figure 3). These higher proportions of higher level student discourse

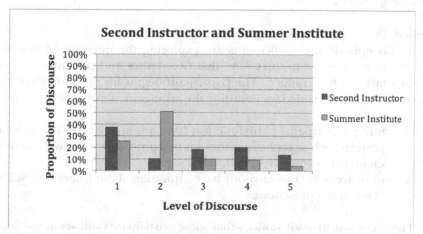

**FIGURE 3. Level of discourse comparisons between summer institute class and second instructor episodes.**

provided initial evidence of the positive impact of the second instructor on student mathematical discourse.

Motivated by the increased high level discourse, the second instructor's interjections into whole-class discussions observed in these sessions were further analyzed and grouped according to their purpose and effect. The following classifications were developed to identify these four types of second instructor interjections.

1.  *Side-trips* are instances when the second instructor pursued an opportunity to discuss an important mathematical idea related to but not necessarily the focus of the lesson.
2.  *Pressing for clarification* occurs when the second instructor detected disequilibrium or confusion and encouraged more thorough explanation of the ideas being discussed.
3.  An *extension* occurs when the second instructor prompted additional exploration and discussion in order to deepen the mathematical understanding of the primary lesson focus.
4.  *Highlighting* serves to bring attention to a student's contribution that might otherwise have gone unnoticed.

All types served to promote meaningful student discourse.

## Episodes Demonstrating High-Level Student Discourse

It was observed that pressing for clarification and extensions provided the greatest opportunity for high level discourse as demonstrated in the following episodes.

### Episode One

In this episode the participants are exploring the question of how to count the number of distinct ways that five dashes and two lines (seven items total) can be arranged. The participant begins his presentation with the seven factorial needed for counting the arrangements of all seven items.

Presenting participant: (A)nytime you have a number of elements in a pattern such as repeating lines (in this problem) you have …seven elements so you have a factorial 7…

Second Instructor: Does anybody have a question about where he got 7!? What is it representing?

The second instructor notices that some participants still seem unclear about the concept and presses the presenting participant for clarification. The ensuing discussion, although it takes a few prompts to move from the level of responding to the second instructor's queries, moves on to a higher level of discourse. Other participants gain access to the problem and help the presenter clarify by relating to previously explored problems.

Second Instructor: So how do you see that? How do you see that was 7!? Can you be specific about where you saw that in the 7 things?

Presenting participant: OK, if you label . . .(the) dash(es) would be one through five, line one and line two. You can arrange each one of those seven different ways. There is a factorial of 7. Does that make sense?

Respondent A: I am wondering does that work only if you are given the exact things that must be in those seven like you must have five dashes and you must have two lines? Would it work for the alphabet, say take the letters of the alphabet and stick any 7 letters in?

Respondent B: Are you asking if you only put them in seven spots like there are only 7 spots but you are using the whole alphabet?

. . .

Respondent C: That would be more like the fifteen books and the three slots.

Respondent D: Or (when) there were nine toppings but we only chose three.

Respondent A: I am still not clear on the 7! . . . Where it has been coming from before was the idea that for the first slot you have 7 choices, for the second slot you have 6 choices . . . this problem (with the 5 dashes and two lines) is different because when I go to choose what I am going to put in the first slot I have two choices.

Presenting participant: That is exactly the problem I had. I couldn't get around that. But you actually have 7 different choices. You can put dash one there or dash two there, they may look the same, but. . .

Respondent A: Thanks, now I see it!

The second instructor's attention to the disequilibrium in the classroom prompts her to press the participant for a more thorough explanation. When participants begin relating (Level Four discourse) the discussion to previously explored problems, the presenting participant is able to make the connection and justify (Level Four discourse) his solution. Because of the second instructor's interjection, higher-level discourse occurs, but more importantly, participants' confusion is resolved. Resolving this confusion allows the discussion to progress to a complete solution of the problem.

*Episode Two*

In this episode the second instructor initiates an extension by asking a group to think about how to solve the pizza problem with the multiplication principle (they had a solution with the addition principle). During their group presentation they mentioned this task. The second instructor clarifies what she had asked them to do and instigates an investigation in table groups by the whole class.

Second Instructor: So the question I asked them, remember yesterday Kathy talked about strategy for when to add and when to multiply depending on how you wrote the problem down. So everyone that I saw, and correct me if I am wrong, kind of thought of this as 0 toppings, 1 topping, 2 toppings. So, you figured those out independently and added them together. Right? So, you kind of separated them into these disjoint mutually-exclusive groups, right, and added them together. So, multiplication was invoked when you kind of built a pizza by making choices, a string of choices. So, what I asked them is could you change your perspective could you look at this problem differently through a multiplication lens instead of an addition and build that $2^6$.

Following this second instructor interjection, the groups work at their tables. After the table group work, the lead instructor facilitates the ensuing group discussion. Without the earlier actions of the second instructor this episode would not have occurred.

Participant A: The fact that there are six toppings and we ended up with two to the sixth makes me think that when we don't have the base piece, like the ice cream or the 39 types of ice cream, that thing that you are loading it on is one. That's what I was trying to see. Is … if I had one pizza … the base pizza is one and I had four toppings I am thinking that the answer will be two to the fourth … on our Pascal's triangle … (the) 5th row down and so I was going to see would that be right if I had four toppings. And it would be.

Lead Instructor: What are you guys think(ing)?

Participant B: I was kind of thinking the same way as Participant A was. I was thinking about the base as one … I was trying to work from the top down (of Pascal's Triangle). And so if I had the one topping not the one at the top but the two.

Lead Instructor: Come up. Do you want to point (to Pascal's Triangle) while you are talking?

Participant B: … we were discussing prior to this that this was the 2 to the 1. Right? And this one up here was 2 to the 0. And so, it was one. So, if this was a one-topping pizza and there was two total ways of doing it … it can either be on the pizza or it can be, it can be plain nothing on the pizza. … So then I have two toppings—so I in my mind

I was trying to get to it the same way you are. …There is something with that two times two… So it can either be this is a topping on my pizza or not on my pizza. So that is two ways … And then this (second topping) can be on there, or these (both toppings) can be on there, and so… (participant is motioning for each topping).

Participant A: I was playing with that piece, too.

Participant B: Yeah, there is suppose to be four ways. If this is my pizza and here is my topping on there (puts pen on document camera …) That is one way. This is another way (takes off pen and puts on a different pen). This is the two-topping way (puts both pens on display). And then, the no-topping way. Right? … It is almost there. There is something with that two, that it is or it isn't. I don't know if that helps or not.

…

Participant C: So there is the first topping, you either get it or you don't. Here is the second topping, you either get it or you don't. You either get it or you don't. Make sense? We are going to keep going. The third topping—

Participant A: There are your twos!

Participant C: Yeah those are my twos. I was so jazzed to see them. You either get it or you don't, you get it or you don't, you get it or you don't. So you make this tree diagram out six layers. It always goes by two's. So you've got every single one.

Participant A: So it's looking at it differently. Like it is either affirmative or it is not. I either want this and do I want to have the next one.

Participant C: So you have got yes or no at every step for every little piece so it is 2 to the 6th. Yep.

The extension introduced by the second instructor allows the participants to justify (Level Four discourse) their approach, using the multiplication principle rather than the addition principle. As they share the solution and the justification, they are able to extend to a generalization (Level Five discourse). The interjection of the second instructor leads the participants to a deepened understanding of the multiplication principle.

In these two episodes the second instructor attends to the disequilibrium in the classroom and introduces an extension. In both cases the student mathematical discourse moves to Level Four including relating and justifying. In fact, discourse reaches the highest cognitive level in the second episode when participants generalize their results. While there are other instances of high level discourse in these classes, the actions of the second instructor clearly play a role in moving the discourse to a higher cognitive level.

## CONCLUSION

The examination of the student mathematical discourse in regular class-rooms and the summer institute classrooms of the grant project revealed distinctly different proportions of cognitive levels of discourse. The regular classes exhibited a decreasing proportion of discourse as the cognitive level increased, a trend similar to that seen in the classrooms of the teacher participants. In contrast, the summer institute classes displayed a higher proportion of discourse at the higher cognitive levels. We attributed at least some of these differences to the involvement of a second instructor in the summer institute classrooms.

The actions of the second instructor were identified as providing additional monitoring of small groups, time management, selecting and sequencing, and interjections into whole class discussion. Four classifications were developed for interjections into whole class discussion: side trips, pressing for clarification, extension, and highlighting. Higher proportions of Levela Three, Four, and Five discourse than seen in other situations occurred following interjections by the second instructor. Pressing for clarification and extensions by the second instructor provided the greatest opportunity for high level discourse.

One of the values of team teaching in this setting was an increased proportion of high-level student mathematical discourse. Episodes when the second instructor acted to press for clarification or extend a discussion displayed the greatest increase. Opportunities for these types of moves may be more likely to be observed by the second instructor than by the lead instructor in part because the lead instructor is focused on the structure and flow of the entire lesson, while the second instructor is able to give full attention to the students and their questions, understanding and disequilibrium. Often the second instructor observed something that the lead instructor did not notice.

The opportunity to participate in this team-teaching structure can provide valuable experience in acting in these second instructor roles. Practice in detecting opportunities for high level discourse, especially pressing for clarification and extensions, can develop an instructor's ability to notice and act on these opportunities when instructing "in solo." Further study is required to determine whether the skills gained through team teaching can be carried over to solo instruction. Additional observations of the subjects in the future could be undertaken to determine whether these instructors who experienced the benefits of team teaching are able to improve their abilities to detect similar opportunities for high level discourse in their solo classrooms.

## REFERENCES

Cook, L., & Friend, M. (1996). Coteaching: Guidelines for creating effective practice. In Meyen, E. L., Vergason, G. A., & Whelan, R. J. (Eds.), *Strategies for teaching exceptional children in inclusive settings* (pp. 152–182). Denere, OH: Love.

Clement, L. (1997). If they're talking, they're learning? Teacher's interpretations of meaningful mathematical discourse. Paper presented at the Annual Meeting of the American Educations Research Association. Chicago, IL.

Dick, T. (2004). Design principles for the Oregon Mathematics Leadership Institute, unpublished announcement.

Grassl, R., & Mingus, T. T. Y. (2007). Team teaching and cooperative groups in abstract algebra: nurturing a new generation of confident mathematics teachers. *International Journal of Mathematical Education in Science and Technology, 38*(5), 581–597.

National Council of Teachers of Mathematics, Commission on Standards for School Mathematics. (1989). *Curriculum and evaluation standards for school mathematics.* Reston, VA: Author

National Council of Teachers of Mathematics, Commission on Teaching Standards for School Mathematics. (1991). *Professional standards for teaching mathematics.* Reston, VA: Author

National Council of Teachers of Mathematics, Standards 2000 Project. (2000). *Principles and standards for school mathematics.* Reston, VA: Author

National Council of Teachers of Mathematics, Commission on Teaching Standards for School Mathematics. (2003). *A research companion to principles and standards for school mathematics.* Reston, VA: Author

Weaver, D. & Dick, T. (2006). Assessing the quality and quantity of student discourse in mathematics classrooms: Year 1 results. http://hub.mspnet.org/media/data/WeaverDick.pdf?media_0000000002280.pdf

Weaver, D. (2007). *Oregon Mathematics Leadership Institute Spring 2007 Evaluation Report.* RMC Research Corporation, Portland, OR.

Yackel, E. (2001). Explanation, justification, and argumentation in mathematics classrooms. *Proceedings of the Twenty-fifth Conference of the International Group for the Psychology of Mathematics Education, 1,* 9–24. Utrecht, The Netherlands.

CHAPTER 2

# CREATING A DISCOURSE-RICH CLASSROOM (DRC) ON THE CONCEPT OF LIMITS IN CALCULUS

## Initiating Shifts in Discourse to Promote Reflective Abstraction

**Robert W. Cappetta and Alan Zollman**

This article describes a quasi-experimental study to create a discourse-rich classroom (DRC). In this article, we discuss the background of reflective abstraction, the argument for initiating shifts in discourse, and the connection from reflective discourse to conceptual understanding. In our study, we designed initiates (individual, peer, instructor, and curriculum) to promote reflective discourse. Initiates were designed for each reflective abstraction construct (interiorization, coordination, generalization, encapsulation, and reversal) in a unit on limits in college calculus. Results show that initiates can promote reflective discourse and conceptual understanding.

**KEYWORDS**: discourse-rich classroom, reflective abstraction, initiates, reflective discourse

*The Role of Mathematics Discourse in Producing Leaders of Discourse*, pages 17–39
**17**

## INTRODUCTION

How does a teacher create a discourse-rich classroom? This article demonstrates how a discourse-rich classroom (DRC) is created for a unit on the concept of limits in calculus, based upon the theoretical background of Piaget, Sfard, Cobb et al., Dubinsky, and Krussel et al. Specifically, this DRC is based upon our assertion that the individual, peer, instructor, and curriculum, can initiate shifts in discourse that may lead to reflective abstraction and promote conceptual understanding of mathematics content.

In this article, we briefly discuss the background of reflective abstraction, the argument for initiating shifts in discourse, and the connection of reflective discourse to conceptual understanding. Next, we describe the design of the DRC to promote reflective abstraction through individual, peer, instructor, and curriculum initiates. We give examples how each initiate (individual, peer, instructor, and curriculum) was designed for each of Piaget's reflective abstraction constructs (interiorization, coordination, generalization, encapsulation, and reversal) in a unit on limits in calculus. We show our DRC results with a quasi-experimental study to improve student performance in college calculus by promoting reflective abstraction through initiates. We close this article with discussing the advantages of designing a DRC through a matrix of initiates and constructs.

## BACKGROUND

Piaget (Beth & Piaget, 1966) states that students construct knowledge through the process of reflective abstraction. According to Piaget, reflective abstraction is present at the earliest stages of cognitive development, and this process continues throughout advanced mathematics. In fact, the development of modern mathematics from primitive mathematics can be viewed as a process of reflective abstraction (Piaget, 1985). One possible way to inspire students to engage in reflective abstraction is through reflective discourse in the classroom. Several authors discuss the importance of discourse in learning mathematics (Cobb, Boufi, McClain, & Whitenack, 1997; Krussel, Springer, & Edwards, 2004; Sfard, 2000).

Sfard (2000) claims communication is the key to understanding mathematical concepts and the need to communicate ideas is the reason individuals construct mathematical objects. She argues that these objects arise when individuals debate mathematical metaphors versus mathematical rigor. She concludes by stating the development of modern mathematics can be viewed as a continuing debate between the metaphor and rigor. Sfard believes that individual thinking behaves much like communication.

She writes,

> Thinking, like conversation between two people, involves turn-taking, asking questions and giving answers, and building each new utterance—whether audible or silent, whether in words or in symbols—on previous ones that are all connected in an essential way (p. 299).

She argues that an examination of public discourse will yield information about thinking, learning and problem solving.

Cobb et al. (1997) examine classroom discourse and Krussel et al. (2004) develop a framework for describing a teacher's participation in the discourse. Cobb et al. (1997) discuss the role of reflective discourse and collective reflection in the classroom. In particular they examine the relationships between discourse and understanding. The authors write that reflective discourse "is characterized by repeated shifts such that what the students and teacher do in action subsequently becomes an explicit object of discussion" (p. 258). They also define collective reflection as "the joint or communal activity of making what was previously done in action an object of reflection" (p. 258). Thus these activities may be crucial to students bringing mathematical objects into being.

Cobb et al. (1997) recognize that all students do not perform the same way in a classroom. They write, "This type of discourse constitutes conditions for the possibility of mathematical learning, but it does not inevitably result in each child reorganizing his or her mathematical activity" (p. 264). The authors recognize that the relationship between concept development and reflective discourse is uncertain. Classroom discourse may help a student construct mathematical concepts and it might accustom a student to mathematical activity. In particular, children may develop a mathematical disposition, as discussed by the National Council of Teachers of Mathematics (1991) by participating in reflective classroom discourse (Cobb et al., 1997).

Cobb et al. (1997) claim that the teacher's role is to "guide, and as necessary, initiate shifts in the discourse such that what was previously done in action can become an explicit topic of conversation" (p. 269). The teacher must invite the students to reflect on their work. She must assess the abilities of the students to participate in the discourse. She may develop a set of symbols used to describe the process being studied (Cobb et al., 1997). These symbols may be a key for many students to develop the conceptual understanding (Gray & Tall, 1994; Sfard, 1991; von Glaserfeld, 1991).

Krussel, Springer, and Edwards (2004) develop a framework for analyzing the teacher's discourse moves. The framework includes purposes, categories, and consequences of discourse moves. The authors define a teacher's discourse moves as the actions of a teacher deliberately designed to affect the discourse in a classroom. Moves include participating in the discourse, mediating the discourse and influencing the discourse. They include several purposes for teachers' discourse moves. These are establishing or changing the environment for the discourse, setting or changing the focus of the discourse, encouraging reflection, requesting justification, encouraging and managing participation, and creating or eliminating disequilibrium.

Krussel et al. (2004) borrow from Heaton (2000) in describing three categories of teachers' discourse moves: scripted, provisional, and improvisational. The authors define scripted moves as a carefully planned collection of moves that the teacher will implement in sequence, provisional moves as a plan of moves to be executed if and when students reach certain benchmarks and improvisational moves as the collection of moves that a teacher uses to react to unforeseen events in the discourse. They argue that the teacher must possess sufficient content knowledge and pedagogical content knowledge to effectively implement improvisational moves.

Krussel et al. (2004) describe several consequences of teachers' discourse moves. These include a shift in the cognitive level of an activity, a shift in the focus of discourse, a shift in authority between the teacher and the group, and an evolutionary development of classroom norms for discourse. They discuss the unintended consequence of a reduction in cognitive level. They examine the work of Stein, Smith, Henningsen, and Silver (2000) and conclude a teacher's activities can dramatically affect the cognitive level of a task.

According to Piaget, et al. (1977). reflective abstraction is a thought process that occurs within an individual. Since it is impossible to know exactly what happens in the mind of another individual, reflective abstraction must be inferred based on evidence. Piaget (Beth & Piaget, 1966) describes four constructs of reflective abstraction. These are interiorization, coordination, encapsulation, and generalization. Piaget describes the importance of reversal in cognitive development (Piaget, Inhelder & Szeminska, 1960). Dubinsky (1991) refines the concept of reversal into a fifth construct and he claims this is essential in advanced mathematics. Descriptions of the five constructs follow.

1. *Interiorization*: Piaget defines interiorization as "translating a succession of material actions into a system of interiorized operations" (Beth & Piaget, 1966, p. 206). Dubinsky (1991) describes interiorization as the construction of internal processes in order to make sense of mathematical concepts. The tools used in constructing these processes include symbols, pictures, and language.

2. *Coordination*: This construction is the process of coordinating two or more processes to obtain a new process (Dubinsky, 1991).

3. *Encapsulation*: Dubinsky and Lewin (1986) state, "Perhaps the most important form of reflective abstraction involves a process of encapsulation" (p. 62). Dubinsky (1991) defines encapsulation as

the conversion of a dynamic process into a static process. Piaget (1985) writes, "Actions or operations become thematized objects of thought or assimilation. . . . The whole of mathematics may therefore be thought of in terms of construction of structures, . . . mathematical entities move from one level to another; an operation on such entities becomes in its turn an object of the theory, and this process is repeated until we reach structures that are alternately structuring or being structured by stronger structures" (p. 49).

4. *Generalization*: Generalization occurs when a student applies an existing schema to a wider collection of concepts. Dubinsky and Lewin (1986) describe the relationship between generalization and encapsulation in the following: "A structure is, in some sense, a form, acting on various aliments as content. After encapsulation this form can become content for other structures which, when generalized, can act upon the encapsulated structure as an aliment" (p. 63).

5. *Reversal*: Piaget does not include reversal as one of his constructs of reflective abstraction, yet he discusses its importance in concept development (Piaget, Inhelder & Szeminska, 1960). Dubinsky (1991) refines Piaget's notion of reversal into a construct of reflective abstraction. Reversal occurs when a student constructs a new mathematical structure by un-doing the processes of a known structure, e.g., viewing subtraction as the inverse of addition.

Recognizing that reflective abstraction is an individual activity, Cobb et al. (1997) claim that the teacher is capable of initiating shifts in the discussion that may lead to reflection. Hershkowitz and Schwarz (1999) state that a rich learning experience promotes reflective processes. Therefore it is our assertion that the individual, peers, the instructor and the curriculum can initiate shifts in discourse that may lead to reflective abstraction and promote conceptual understanding of mathematical concepts.

What is the role of discourse in promoting reflective abstraction? First, Piaget (Beth & Piaget, 1966) claims that reflective abstraction is a personal activity. Second, Cobb, Jaworski, and Presmeg (1996) discuss the relationship between social discourse and individual reflective abstraction. Lampert (1986), as cited in the NCTM *Principles and Standards for School Mathematics* (2000), states that classroom discourse and social interaction are used to assimilate new ideas and to develop connections among them. Third, Cobb et al. (1997) claim that the teacher is capable of initiating shifts in the discourse that may lead to reflective abstraction. Finally, several curricula have been developed in recent years to encourage students to begin to reflect about their thinking in mathematics. Therefore the goal of a discourse rich classroom is to provide initiates that may lead to reflective abstraction.

Teachers, peers and the curriculum cannot guarantee that reflective abstraction occurs. Piaget (Beth & Piaget, 1966) claims reflective abstraction is an individual activity. Despite the best efforts by teachers and curriculum designers, the individual alone is capable of engaging in reflective abstraction, but the teacher may infer reflective abstraction based on student performance.

## THE QUASI-EXPERIMENTAL CURRICULUM

Selden, Mason, and Selden (1989) confirm that calculus students do not understand the fundamental concepts. These students are unable to solve non-routine problems. Their research also suggests that calculus students recall very little calculus in later classes (Selden, Selden, Hauk, & Mason, 1999).

This study examines the concept of limit. It was chosen because it is the first idea in calculus that is substantially different from algebra. Additionally, this idea is central to the notions of derivative and integral that appear later in the curriculum. Therefore a discourse-rich curriculum designed to initiate the constructs of reflective abstraction may improve student understanding of the concept of limit and, thus, increase a student's chances of success in calculus.

A class of 16 students at a community college participated in the project. They spent one week studying the concept of limit. Class sessions included minimal lectures from the teacher, significant group work on problem sets, extensive student writing about the meaning of concepts, and collective reflection at the end of each lesson. Classroom problem sets included exercises that required the use of a graphing calculator. Homework sets consisted of traditional textbook questions assigned from *Calculus* by Larson, Hostetler and Edwards (2005). The unit on limits followed Thompson's (1985) five guiding principles for an effective mathematics curriculum:

1. Be problem based.
2. Promote reflective abstraction.
3. Contain (but not necessarily be limited to) questions that focus on relationships.
4. Have as its objective a cognitive structure that allows one to think with the structure of the subject matter.
5. Allow students to generate feedback from which they can judge the efficacy of their methods of thinking (p. 200).

This curriculum was designed to initiate each of the constructs of reflective abstraction (see Figure 1). Those constructs are interiorization, coordination, generalization, encapsulation and reversal. According to Dubinsky, the individual, peers, instructor or curriculum materials can initiate these

| Contruct <br> Initiate | Interiorization | Coordination | Encapsulation | Generalization | Reversal |
|---|---|---|---|---|---|
| Instructor | Instructor × Interiorization | Instructor × Coordination | Instructor × Encapsulation | Instructor × Generalization | Instructor × Reversal |
| Peer | Peer × Interiorization | Peer × Coordination | Peer × Encapsulation | Peer × Generalization | Peer × Reversal |
| Individual | Individual × Interiorization | Individual × Coordination | Individual × Encapsulation | Individual × Generalization | Individual × Reversal |
| Curriculum | Curriculum × Interiorization | Curriculum × Coordination | Curriculum × Encapsulation | Curriculum × Generalization | Curriculum × Reversal |

**FIGURE 1. Matrix of Constructs × Initiates combinations.**

constructs. The matrix in Figure 1 includes twenty combinations of construct × initiate. A collection of observed classroom examples follows in the next section.

## CLASSROOM EXAMPLES OF INSTRUCTOR INITIATES TO PROMOTE REFLECTIVE ABSTRACTION

### 1. Instructor × Interiorization

The instructor promoted interiorization with a brief lecture on the algebraic procedure evaluating limits. During this lecture she focused on the procedural elements such as adding rational expressions, rationalizing the denominator and substitution. After her lecture she asked the students to complete a few of these problems. She circulated through the room and checked progress.

**Example:**
After students finished the problems, the instructor asked the students to describe, procedurally, how they solved the problems. Assisting students with procedural understanding is an instructor's initiate of interiorization.

While students worked on the problems together, the instructor circulated throughout the room, asked questions and provided hints. During this time she initiated coordination, generalization, encapsulation and reversal. These initiates are categorized as follows.

### 2. Instructor × Coordination

The following are observed examples of the instructor's initiates of coordination, provided by the experimental unit.

*Example 1:*

If a function is heading to positive infinity, does the limit exist? (Graphical representation of a function, notion of infinity and concept of limit are coordinated.)

*Example 2:*

If the graph is connected and the table tells us that the limit exists, how does this relate to the definition of continuity? (Graphical representation of a function, tabular representation of a function, concept of continuity and concept of a limit are coordinated.)

*Example 3:*

How is the table related to the graph for a function with an asymptote? (Graphical representation and tabular representation of a function are coordinated.)

*Example 4:*

How is the notion of removable discontinuity related to the division by zero issue? (Removable discontinuity, division by zero and concept of a limit are coordinated.)

## 3. Instructor × Encapsulation

The following are observed examples of the instructor's initiates of encapsulation, provided by the experimental unit.

*Example 1:*

After looking at the formal definition of limit, what does it mean to you? (The instructor encouraged the student to develop a personal definition of limit—one that was mathematically correct and applicable to other contexts.)

*Example 2:*

What does it *really* mean for a function to be continuous? (The instructor encouraged the student to develop a personal definition of continuity—one that was mathematically correct and applicable to other contexts.)

## 4. Instructor × Generalization

The following are observed examples of the instructor's initiates of generalization, provided by the experimental unit.

*Example 1:*
How can one decide that a piecewise function is continuous?

*Example 2:*
Can a table make it look like a limit exists but it really doesn't?

*Example 3:*
How is this problem similar or different from the previous problem?

*Example 4:*
In light of what you now know, can you go back and solve the previous problem using a different strategy?

At the conclusion of the lessons she would address the entire class and ask clarifying questions. These questions served as collective reflection and reversal and encapsulation questions were prominent.

## 5. Instructor × Reversal

The following are observed examples of the instructor's initiates of reversal, provided by the experimental unit.

*Example 1:*
If the function does not exist at a point, what does the graph look like?

*Example 2:*
If a function is not continuous, can a limit exist?

*Example 3:*
If a limit does not exist, what would the table look like?

The instructor successfully provided initiates for interiorization, coordination, generalization, reversal and encapsulation. She did this by clarifying procedures and establishing relationships among concepts. She was comfortable with initiating interiorization and coordination, but the relatively small number of questions in the other areas indicates that she was less comfortable initiating generalization, reversal, and encapsulation.

## CLASSROOM EXAMPLES OF PEER INITIATES TO PROMOTE REFLECTIVE ABSTRACTION

In order to facilitate communication between peers, the classroom problems referred to a hypothetical study group consisting of Alice, Tom,

George, and Carla. Alice uses algebraic procedures to solve problems; Tom uses tables; George uses graphs; Carla connects ideas. Students were asked to discuss how these hypothetical students would solve a certain problem. In order to help students coordinate ideas, they eventually had to write how Carla would solve a problem. Several examples of peer initiates of reflective abstraction were identified. The examples that follow are a sample of questions that students asked while working in collaborative groups. The questions indicate initiates of interiorization, coordination, and generalization

## 1. Peer × Interiorization

There were several examples of students (as peers) executing procedures. The working student would explain how she was solving the problem. When other students were unclear, they would ask questions about how a procedure is executed. Interiorization inferences follow.

**Example 1:**
Use a tabular procedure to find the limit.

Student 1: How do you solve this with the table?
Student 2: Plug in the numbers to find out what the values are.
Student 1: How does that help us find the limit?
Student 2: Where are the numbers going? What are they getting closer and closer to?

**Example 2:**
Use an algebraic procedure to find the limit.

Student 1: When you plug in 0 you get $0 - 2$ which is negative 2, right. So that's your $f$ of 0, right, that's a negative 2 and when you look at the graph, it's a straight line sense.

## 2. Peer × Coordination

The largest number of peer initiates was for the coordination category. Students tried to understand the various concepts and they wanted to see how these ideas were related. A collection of student coordination discussions follows.

*Example 1:*

Coordinate graphic and algebraic representations.

Student 1: There is no fraction, so can there be a hole?
Student 2: There is nothing to cancel out so there is no hole.

*Example 2:*

Coordinate several strategies to find a solution.

Student 1: What should I write for Carla?
Student 2: She looks at all three strategies.
Student 3: If you get the same answer for each, you have three ways to
verify that your answer is correct.

*Example 3:*

Coordinate several strategies to find a solution.

Student 1: Wait a minute, should there be a hole there?
Student 2: Yes.
Student 1: But why can't we see it on the calculator?
Student 2: Because of how the *x*-axis is defined.
Student 3: Should Carla invalidate George's method because we didn't
even see the hole in the graph? So the graph method might not be
that good for this one.
Student 2: But it still would have a limit though.
Student 2: It is the same thing no matter what way you do it.
Student 3: But for convenience factor, George's would not be the most
convenient for this set up and Tom's, it doesn't exactly say 0.5 but
you can infer from the information that the limit will approach 0.5.
Student 1: Because there is the same change every time.
Student 3: So you probably want to use Alice's first and Tom's to verify.
Student 1: Yeah.
Student 2: I think you should always try algebraically first off.

*3. Peer × Encapsulation*

No peer by encapsulation examples were observed.

## 4. Peer × Generalization

There were relatively few observed examples of peer initiates of generalization.

***Example:***
Construct a graph with certain characteristics.

Student 1: We are basically doing the opposite of the last one.
Student 2: How do we do that? Just put a negative sign in front.
Student 3: I have the graph.
Student 2: Good job.
Student 1: Construct a graph . . .
Student 2: Is there some way that we can manipulate this to make it as given?
Student 1: I don't know. I forget how to make equations like that. . . .
Student 2: No. It didn't work.
Student 3: Try one over one minus $x$.
Student 1: No. There needs to be a zero.
Student 2: That's hard.
Student 3: You have to cube it. It is $(x+2)$ over $(x-1)$. Cool.

## 5. Peer × Reversal

No examples of reversal were inferred among peer initiates in the experimental curriculum. The lack of student initiates in encapsulation or reversal may be related to the relatively small number of instructor initiates in these areas.

## CLASSROOM EXAMPLES OF CURRICULUM INITIATES OF REFLECTIVE ABSTRACTION

The curriculum was designed to initiate interiorization, coordination, generalization, reversal, and encapsulation with respect to the notion of limit. There were a large number of questions in most categories. A sample of these questions together with student answers follows.

## 1. Curriculum × Interiorization

Many questions in the designed curriculum ask students to complete procedures. These include constructing tables to suggest values of a limit, evaluating a limit algebraically, and determining whether or not a function is continuous at a given point. A sample of interiorization initiates together with actual student responses follows.

*Example 1:*
　Use a graphical procedure to find a limit.

　Designed-Curriculum Question: Carla asks George (the hypothetical, designed-curriculum students) to explain how the graph shows the function approaching the same value as $x$ approaches 2 from the left and the right. How will George answer Carla's question?
　Student Response: The graph shows the line from the left and the right approaching $x = 2$ and the $y$-value is approaching $y = 2$.

*Example 2:*
　Use an algebraic procedure to find a limit.

　Designed-Curriculum Question: Alice likes algebraic simplification. She claims in this case it is appropriate to plug 2 into $g(x)$ in order to determine the behavior of the function as $x$ approaches 2 from the left and from the right. How would she answer the question? Demonstrate the strategy.
　Student Response:

$$\frac{x^2 - 4x - 5}{x+1}, \quad \frac{2^2 - 4 \cdot 2 - 5}{2+1} = \frac{4 - 8 - 5}{3} = \frac{-9}{3} = -3.$$

## 2. Curriculum × Coordination

　Many questions were asked to initiate coordination. The curriculum was designed to help students understand limits from a tabular, graphical, and algebraic perspective. The students also were expected to understand how notions like asymptotes, removable discontinuities, and one-sided limits were related to the notion of limit. For these reasons the largest number of questions related to initiating coordination. A sample of curriculum questions and student answers follows.

*Example 1:*
　Coordinate notions of limit and continuity.

Designed-Curriculum Question: Will a function always approach the same number from both the left and the right? Write a paragraph. Include examples and counterexamples in your discussion. Discuss how this idea is related to other ideas in the unit.

Student Response: A function will not always approach the same number from both the left and the right. Cases where a function approaches the same number from the left and from the right include functions that are always continuous. There are functions that are not continuous but still approach the same number from the left and the right. In the graphs of this type of function there is a hole.

*Example 2:*
Coordinate algebraic solution and tabular solution.

Designed-Curriculum Question:
Let

$$f(x) = \frac{x^2 - 4}{x - 2}.$$

Evaluate

$$\lim_{x \to 2} f(x).$$

Student Response: Student solves the problem two ways. The first uses the algebraic strategy recognizing that the function $x + 2$ agrees with the original function at all but one point, and then she substitutes in 2 to get an answer of 4. She also constructs an $x$, $y$ table with $x$-values of 1.997, 1.998, 1.999, 2, 2.001, 2,002, 2.003, and even though the function is not defined at 2, she concludes the limit is 4.

*Example 3:*
Coordinate the notions of limit and continuity.

Designed-Curriculum Question: Explain the relationship between the concept of a limit and the notion of a continuous graph.

Student Response: The value of the limit and the value of the function, at the same $x$-value, must correspond for a graph to be continuous. Therefore it can be stated that if the limit is equal to the function value then it can be assumed that the function is continuous at that point.

## 3. Curriculum × Encapsulation

There was one primary question in the curriculum that asked students to encapsulate their understanding of the concept of limit. Two examples of student work follow:

**Example:**

Encapsulate the notion of limit.

Designed-Curriculum Question: Carla decides to write a summary of this collection of limit lessons in her notebook. She wants to write a definition in her own words and she wants to include relevant examples and counterexamples in her notes. Help Carla complete her task.

Student Response: A limit is when, as the function is getting closer and closer to the same *x*-value from the left and the right, the function is getting closer and closer to the same *y*-value from the left and from the right. Continuous graphs always have a limit for any *x*-value. You can draw a continuous graph without lifting your pencil from the paper. For a graph that is not continuous, you have to lift your pencil from the paper to keep drawing it. Functions that have a limit even though they are not continuous reach the same number from the left and from the right even if there may or may not be a *y*-value at that point. Examples include functions with removable discontinuities that create holes in graphs. You can use the all-but-one-point rule to find the limit. A function that is not continuous and has no limit is because the function gets closer to a different number from the left and from the right. You can use tables or graphs to see this. The

$$\lim_{x \to c} f(x) = L$$

means when $x$ approaches c from both the right and the left side,

$$f(x) = L.$$

So,

$$\lim_{x \to c^-} f(x) = L \text{ and } \lim_{x \to c^+} f(x) = L, \text{ then } \lim_{x \to c} f(x) = L \cdot$$

When $f(x)$ approaches different numbers from left and right, the limit does not exist. When $f(x)$ increases or decreases without bound as $x$ approaches $c$, the limit does not exist.

Ex.:

$$\lim_{x \to 0} \frac{1}{x^2}.$$

There is a relationship between limits and continuous graphs. If

$$\lim_{x \to c} f(x) \neq f(c)$$

then we can conclude that function $f$ is not continuous at $x = c$ so the limit exists everywhere except at $x = c$, I also realized that existence or nonexistence of $f(x)$ at $x = c$ has no bearing on the existence of the limit of $f(x)$ as $x$ approaches $c$.

$$\text{Ex. } \lim_{x \to 0} \frac{x}{\sqrt{x+1} - 1} = 2.$$

## 4. Curriculum × Generalization

A few questions in the experimental curriculum were designed to help students generalize their ideas into other areas. Many of these questions asked students to hypothesize rules. Other questions asked students to construct functions with certain characteristics. There were relatively few generalization questions. A sample of these generalization initiates together with student responses follows.

**Example 1:**
Construct a graph with specific characteristics.

Designed-Curriculum Question: The instructor asks the team to construct a function $q(x)$ and its graph such that both of the following statements are true: As $x$ gets closer and closer to 3 from the left, $q(x)$ gets increasingly positive without bound. As $x$ gets closer and closer to 3 from the right, $q(x)$ gets increasingly positive without bound.
Student Response: Student constructs the rule

$$q(x) = \frac{4}{(x-3)^2}$$

and she constructs the graph of the function.

**Example 2:**
Extending the "division by zero" case to permit a limit.

Designed-Curriculum Question: Explain whether or not it is possible for a limit to exist if a "zero in the denominator" results after plugging in the appropriate value.

Student Response: There is a possibility for it to exist if plugging in the appropriate value also causes a zero in the numerator of the function.

## 5. Curriculum × Reversal

The curriculum initiated reversal by asking students to reverse definitions or to construct counterexamples. A sample of reversal questions and student definitions follows:

### Example 1:
Reverse the definition of the limit.

Designed-Curriculum Question: Construct a graph to show when a limit does not exist.

Student Response: If there is an asymptote at $x = -1$, then the limit does not exist at $x = -1$.

### Example 2: Reverse the definition of the limit.

Designed-Curriculum Question: Is it possible for a limit to exist if a "zero in the denominator" results after "plugging in" the appropriate value?

Student Response: A special case happens when an $x$-value gives you a zero in both the numerator and denominator. You can't say that $5/0 = 0$ because $0 \times 0 \neq 5$, but $0/0$ may equal any number because any number $\times 0 = 0$. When this occurs you may be able to simplify the function to find a limit. In the function

$$f(x) = \frac{x^2 + x + 6}{x - 2},$$

you can't use $x = 2$ because it gives a zero in the denominator. But if you plug in 2 in the numerator you will notice that it also gives a zero. Because of that reason the numerator should be examined to see if it can be factored and it can. $x^2 + x - 6$ can be factored into $(x - 2)(x + 3)$. Since there is also an $(x - 2)$ on the bottom, the two can be canceled. You are left with only $(x + 3)$. If you compare the graphs of

$$f(x) = \frac{x^2 + x + 6}{x - 2} \text{ and } x + 3,$$

you will notice that they both agree at all points except at $x = 2$. The limit exists at $x = 2$ because the function approaches 5 from both the left and the right.

## CLASSROOM EXAMPLES OF INDIVIDUAL INITIATES TO PROMOTE REFLECTIVE ABSTRACTION

Students were assigned textbook questions from Larson, Hostetler, and Edwards (2005) *Calculus* for homework. This work was analyzed to find evidence of interiorization, coordination, generalization, encapsulation, and reversal. These examples of reflective abstraction occurred while students completed homework independently. Unlike the curriculum, these problems did not prompt students to reflect on their work; for this reason they are classified as individual initiates.

### 1. Individual × Interiorization

Students regularly performed the procedures needed to solve the problems. They evaluated many limits using algebraic procedures. Procedural knowledge is a standard requirement in most textbook exercise sets, so interiorization was the most prevalent construct in the homework.

### 2. Individual × Coordination

Students engaged in coordination on the textbook homework. Students would often use more than one strategy to evaluate the limits. They would use algebra, tables and graphs. The students also successfully coordinated notions of one-sided limits, limits, asymptotes and continuity. The following are a few examples of coordination from the homework.

**Example 1:**
Coordinate limit and continuity.

The limit exists and the function value exists, but they are not equal. The graph is not continuous, and there is a hole.

*Example 2:*
   Coordinate removable discontinuity and the all-but-one-point rule.

The limit at $f(5)$ is $1/10$. Substitution doesn't work initially. Once the hole is "removed," we can find the limit of the function based on the all-but-one-point rule.

*Example 3:*
   Coordinate limit and continuity.

The value of the function at $c$ doesn't match the limit of the graph at $c$, so the function is not continuous.

## 3. Individual × Encapsulation

There were no opportunities in the traditional homework for students to demonstrate encapsulation.

## 4. Individual × Generalization

There were few opportunities to demonstrate generalization of the limit concept. One such question was, "Is

$$\lim_{x \to 0} \sqrt{x} = 0$$

true?" Students had to decide how to extend the definition of the limit to this case. Some students recognized that it was impossible to approach 0 from the left, so they said as $x$ approaches 0, $f(x)$ tends to 0, so the limit exists. Others said it was false because it was impossible to approach 0 from the left.

## 5. Individual × Reversal

For certain true-false questions, students needed to demonstrate why a statement was false. For example students had to reverse the limit definition to explain why

$$\lim_{x \to 0} \frac{|x|}{x} = 1$$

is a false statement. A student says as $x$ approaches 0 from the left, the value of the function always is $-1$, and as $x$ approaches 0 from the right, the value of the function always is +1; therefore the limit does not exist and the statement must be false.

## RESULTS

This discourse-rich classroom was effective in promoting reflective abstraction. Some construct-initiate pairs were very prevalent and others appeared rarely if at all. Figure 2 indicates how often pairs occurred.

These results show that the constructs of interiorization and coordination were most prevalent. Generalization and reversal occurred much less often. Encapsulation occurred rarely. These results indicate that students, peers and teachers should be encouraged to initiate generalization, coordination and reversal. Since this does not occur spontaneously training in initiating these constructs may be necessary.

Cappetta (2007) shows that a classroom designed to promote reflective abstraction improves student performance in calculus and improves student ability to communicate mathematics in writing. However, a classroom such as this is not effective for all students. Several students in the project did not perform well. They were given several opportunities to engage in reflective abstraction yet many ignored assignments or provided terse responses. A few of these students indicated displeasure with the project and commented that they should not be required to write about mathematics or explain their reasoning. Their negative attitudes toward the project together with their lack of interest in reflective abstraction may have been contributing factors to their lack of success. In future research, working on classroom norms and managing student expectations may improve these attitudes.

| Construct Initiate | Interiorization | Coordination | Encapsulation | Generalization | Reversal |
|---|---|---|---|---|---|
| Instructor | often | often | never | rarely | rarely |
| Peer | often | often | never | rarely | never |
| Individual | often | often | rarely | often | rarely |
| Curriculum | often | often | rarely | often | often |

**FIGURE 2. Instances of Initiate × Construct Classroom Observations**

## CLOSURE

In our Discourse-Rich Classroom (DRC) the evidence *implies* that our design was successful in initiating reflective abstraction through instructor, peer, curricular, and individual initiates. Evidence of interiorization and coordination was seen most often. This was especially true with the peer initiates and the instructor initiates. Evidence for generalization and encapsulation was rare or nonexistent when students worked on the problems together or individually. In order to increase opportunities for reflective abstraction, teachers should be encouraged to promote generalization, reversal, and encapsulation. Textbook problems also may be augmented with additional questions to initiate each of the five constructs. Students working together should be encouraged to ask generalization, reversal and encapsulation questions. Since we found that most students engage in mostly interiorization and coordination, these recommendations should enable students to have more diverse opportunities to engage in reflective abstraction.

However, as Cobb et al. (1997) discuss, a discourse-rich classroom provides the opportunity for learning; it does not necessarily cause learning. Piaget (Beth & Piaget, 1966) says reflective abstraction is an individual activity. Teachers, peers, or a curriculum may try to initiate reflective abstraction, but whether or not it occurs depends on the individual student.

Why does one student learn, while another student in the same class does not? Understanding the mathematical content occurs when the student decodes, conceptualizes, and applies the content—when reflection occurs. Our goal was to demonstrate the characteristics of a DRC. We have shown that discourse initiated through individual, peers, instructor, and curriculum initiates can promote reflective abstraction. In particular, we argue that discourse-rich activities can promote interiorization, coordination, encapsulation, and reversal. Since reflective abstraction is essential to develop understanding, a discourse-rich classroom is an effective tool for mathematics teaching and mathematics learning.

## REFERENCES

Beth, E., & Piaget, J. (1966). *Mathematical epistemology and psychology*. Dordrecht, The Netherlands: D. Reidel.

Cappetta, R. (2007). Reflective abstraction and the concept of limit: A quasi-experimental study to improve student performance in college calculus by promoting reflective abstraction through individual, peer, instructor and curriculum initiates. Unpublished doctoral dissertation, Northern Illinois University, DeKalb, IL.

Cobb, P., Boufi, A., McClain, K., & Whitenack, J. (1997). Reflective discourse and collective reflection. *Journal for Research in Mathematics Education, 28*(3), 258–277.

Cobb, P., Jaworski, B., & Presmeg, N. (1996). Emergent and sociocultural views of mathematical activity. In Steffe, L., Nesher, P., Cobb, P., Goldin, G., & Greer, B. (Eds.), *Theories of mathematical learning* (pp. 3–19). Mahwah, NJ: Lawrence Erlbaum.

Dubinsky, E. (1991). Reflective abstraction in advanced mathematical thinking. In Tall, D. (Ed.), *Advanced mathematical thinking* (pp. 95–126). Boston: Kluwer.

Dubinsky, E., & Lewin, P. (1986). Reflective abstraction and mathematics education: The genetic decomposition of induction and compactness. *Journal of Mathematical Behavior, 5*(1), 55–92.

Gray, E., & Tall, D. (1994). Duality, ambiguity and flexibility: A proceptual view of simple arithmetic. *Journal for Research in Mathematics Education, 26*(2), 115–141.

Heaton, R. M. (2000). *Teaching mathematics to the new standards.* Reston, VA: National Council of Teachers of Mathematics & Teacher's College Press.

Hershkowitz, R., & Schwarz, B. (1999). Reflective processes in a technology-based mathematical classroom. *Cognition and Instruction, 17,* 66–91.

Krussel, L., Springer, G. T., & Edwards, B. (2004). The teacher's discourse moves: A framework for analyzing discourse in mathematics classrooms. *School Science and Mathematics, 104*(7), 307–312.

Lampert, M. (1986). Arithmetic as problem solving. *Arithmetic Teacher, 36,* 34–36.

Larson, R., Hostetler, R., & Edwards, B. (2005). *Calculus.* New York: Brooks-Cole.

National Council of Teachers of Mathematics. (1991). *Professional standards for teaching mathematics.* Reston, VA: Author.

National Council of Teachers of Mathematics. (2000). *Principles and standards for school mathematics.* Reston, VA: Author.

Piaget, J. (1985). *Equilibration of cognitive structures.* Chicago: University of Chicago Press.

Piaget, J., & collaborators. (1977). *Recherches sur l'abstraction reflechissante. Vol. I & II.* Paris: Presses Universitaire de France.

Piaget, J., Inhelder, B., & Szeminska, A. (1960). *The child's conception of geometry.* New York: Basic Books.

Selden, A., Selden, J., Hauk, S., & Mason, A. (1999). *Do calculus students eventually learn to solve nonroutine problems?* Tennessee Technological University Mathematics Department Technical Report No. 1999-5.

Selden, J., Mason, A., & Selden, A. (1989). Can average calculus students solve nonroutine problems? *Journal of Mathematical Behavior, 8,* 45–50.

Sfard, A. (1991). On the dual nature of mathematical conceptions: Reflections on processes and objects as different sides of the same coin. *Educational Studies in Mathematics, 26,* 61–86.

Sfard, A. (2000). Steering (dis)course between metaphors and rigor: Using focal analysis to investigate an emergence of mathematical objects. *Journal for Research in Mathematics Education, 37*(3), 296–327.

Stein, M. K., Smith, M. S., Henningsen, M. A., & Silver, E.A. (2000). *Implementing standards-based mathematics instruction.* Reston, VA: National Council of Teachers of Mathematics.

Thompson, P. W. (1985). Experience, problem solving, and learning mathematics: Considerations in developing mathematics curricula. In Silver, E. A. (Ed.), *Teaching and learning mathematical problem solving: Multiple research perspectives* (pp. 189–243). Hillsdale, NJ: Erlbaum.

Von Glaserfeld, E. (1991). Abstraction, re-presentation, and reflection: An interpretation of experience and Piaget's approach. In Steffe, L. P. (Ed.), *Epistemological foundations of mathematical experience* (pp. 45–65). New York: Springer-Verlag.

CHAPTER 3

# DISCURSIVE PRACTICES IN COLLEGE PRE-CALCULUS CLASSES

Jo Clay Olson, Libby Knott, and Gina Currie

Many students drop out or do not pass their first college mathematics class. This sometimes causes them to change their major to one that does not require a strong mathematics background. To increase student success rates, we used K–12 research (Cobb, 2000) to investigate the role of class structure on the mathematical learning of undergraduate students. Each K–12 mathematics classroom is a unique community created by the instructor and students (Davis & Simmt, 2003) and is influenced by the class structure. Research on discourse in K–12 mathematics classrooms suggests that specific characteristics of mathematical discourse influence student learning. (Davis & Simmt, 2003; Kazemi & Steipek, 2001; Ladson-Billings, 1995; Sfard, 2008).

This research project sought to adapt some aspects of classroom structure and discourse and investigate the results on students in a college pre-calculus class. Small class size provided opportunities for increased interaction between students and their instructor, and the ensuing discourse had the potential for teachers to assess student needs and acquisition of mathematical knowledge. In comparing student learning in problem-based learning (PBL) environments with traditional content-based instruction, Boaler (1998) found that students who learn through a problem-based approach exhibit higher achievement on standardized tests and on problem solving tests dealing with realistic situations than students from classrooms using a traditional approach. We describe the teacher and student discourse in four pre-calcu-

*The Role of Mathematics Discourse in Producing Leaders of Discourse*, pages 41–59
Copyright © 2009 by Information Age Publishing
All rights of reproduction in any form reserved.

lus classes, and focus on two particular classes, one employing PBL and the other a traditional lecture-based class, in order to compare and contrast the two classes. We attempt to explain the reasons for the differences in student achievement by examining three key components of teaching that influenced student learning in these two classrooms: *teacher beliefs* about their roles and that of their students; the *classroom discourse*; and the *classroom environment.*

**KEYWORDS:** Teacher beliefs; classroom discourse; classroom environment; pre-calculus

## INTRODUCTION

The Mathematics Association of America reported that college freshmen who enroll in pre-calculus classes across the United States are being filtered out of fields like engineering, science, and mathematics (STEM) because of their lack of success in entry-level college mathematics classes. Twenty percent of students intending to major in science or engineering need remediation in mathematics in spite of completing more mathematics courses in high school (Higher Education in Science and Engineering, 2004; Lutzer, Rodi, Kirkman, & Maxwell, 2007). Thus, many students who want to pursue STEM careers find themselves taking a pre-calculus course and approximately 50 percent of them fail (Cohen, 2006). Failure in pre-calculus often translates to dashed dreams and a new career focus. Sometimes, students drop out of college altogether. Thus, a major challenge is how to increase student achievement in pre-calculus.

To address this challenge, we consulted the research and found limited resources regarding the teaching of entry-level college mathematics. Like Rasmussen, Kwon, Allen, Marrongelle, & Burtch (2006), we theorized that research originating from K–12 could inform the teaching and learning of undergraduate mathematics. In particular, the structure of K–12 mathematics classes appears to influence the learning opportunities afforded students (Cobb, 2000). From this literature, three classroom structures emerged that we felt could be implemented with minimal support in college pre-calculus courses. The three class structures were:

1. problem-based learning ([PBL] Kyeong, 2003, Schoenfeld, 1998);
2. student collaboration in small groups ([CSG] Grouws & Cebulla, 2000; Kenny, Kenny, & Dumont, 1995); and
3. smaller class size (Biddle & Berliner, 2002).

Both PBL and CSG suggested a framework for instruction that changed student engagement from listening and note taking to active problem solving. These different actions suggested that discourse among students and between the instructor and students may also change. Small class size provided opportunities for increased interaction between students and their instructor. Thus, discourse had the potential for teachers to assess student needs and acquisition of mathematical knowledge. We hoped to increase student achievement by changing the traditional pre-calculus class through implementing these three alternative classroom structures that emerged from the literature.

The purpose of this study was to describe how these three classroom structures were enacted and how they contrasted with a traditional lecture-style class. In addition, we wanted to identify and describe the opportunities that led to increased student learning. Recognizing that interactions between an instructor and his or her students support students' learning (Davis & Simmt, 2003; Lave, 1996; Sfard, 2008; Turner et al., 2002; Walkerdine, 1997), we focused our research questions on classroom discourse. Specifically, we sought to describe discourse within PBL, CSG, small class size, and traditional classroom structures and use this to show how the presentation of new content impacted student learning opportunities. A brief discussion of the three class structures is followed by a conceptual framework for examining the discourse in four sections of pre-calculus.

## PROBLEM-BASED LEARNING

PBL describes a learning environment where problems drive the learning (Kyeong, 2003). That is, learning begins with a problem to be solved and the problem is posed in such a way that students need to gain new knowledge before they can solve the problem (Erickson, 1999; Hiebert et al., 1996; Hiebert et al., 1997; Krulik & Rudnick, 1999; Lewellen & Mikusa, 1999). Students build their understanding of mathematics and heuristics by successfully using prior experience to solve new problems (Boaler, 1998; Shoenfeld, 1985). In comparing student learning in PBL environments with traditional content-based instruction, Boaler found that students who learn through a problem-based approach exhibit higher achievement on standardized tests and on problem solving tests dealing with realistic situations than students from classrooms using a traditional approach.

## STUDENT COLLABORATION IN SMALL GROUPS

CSG engages students in a common task as they search for understanding, meaning, or solutions (Kenny, Kenny, & Dumont, 1995). During collaboration, students discuss prior experiences that relate to the problem, strategies that could be used to solve the problem, and whether the solution is reasonable. Research (Grouws & Cebulla, 2000) indicates that problem solving in small groups supports learning. Grouws and Cebulla reviewed 80 studies that compared student achievement in two settings: problem solving in CSG and whole-class instruction in which students independently solved problems. They found that 40 percent of the studies indicated that students working in CSGs out-performed students in traditional whole-class learning environments on performance-based assessments. In only 2 of the 80 studies did the students in the traditional whole-class learning environments perform better than the students in CSGs. In the other studies (58%) there were no significant difference in performance. This meta-analysis of the literature suggests that student learning may be positively impacted by CSGs.

## SMALLER CLASS SIZE

Small class size is thought to create space and time for more student participation, individualized instruction, and positive interactions among students. Biddle and Berliner (2002) suggested that small class environments are structurally different from large classes and this structural difference supported learning. The impact of small class size was greatest in the primary grades, and when small classes were maintained in the upper grades the positive impact continued. However, their study also revealed that the impact was minimal when small class size occurred only in the upper grades. In a meta-analysis of empirical studies on class size and student achievement, Glass, Cahen, Smith, and Filby (1982) found that student achievement increased when the instruction was more than 100 hours. These results suggest that small classes in an isolated college course with less than 100 hours of contact time may not actually support student learning. In fact, Jarvis (2007) found that a small class size in college calculus had little effect on student achievement and dropout rates. In spite of these studies, class size continues to be perceived by faculty and society as a contributing factor to low student achievement.

## CONCEPTUAL FRAMEWORK

Each mathematics classroom is a unique community created by the instructor and students (Davis & Simmt, 2003). Social interactions, class structure, and curriculum combine to create opportunities for the unique discourse that may develop in a mathematics classroom. It is however, the type of discursive practices of the teacher that dictate the culture of the classroom (Turner et al, 2002) through implicit and explicit ways (Boaler, 2000). Research on discourse in K–12 mathematics classrooms suggests that specific characteristics of mathematical discourse influence student learning. (Davis & Simmt, 2003; Kazemi & Steipek, 2001; Ladson-Billings, 1995; Sfard, 2008).

Classroom discourse typically follows a three-part exchange beginning with a teacher initiation, followed by a student response, and then the teacher's response (IRE, Cazden, 2001). Teachers who maintain control of the classroom dialogue through this discourse pattern often articulate a belief that the teacher's role is to transmit knowledge to students. However, Truxaw (2009) found that "simply engaging students more actively in classroom discourse is not a panacea for improving mathematical achievement" (p. 18). A grade-eight teacher, Mr. Larson, used triadic exchanges to promote meaning-making by inviting students to hypothesize, justify, and make sense of mathematics. Truxaw theorized that triadic exchanges are not limited to conveying a teacher's ideas, but they can also build students' understanding. This depends on the *function* of the discourse; either univocal or dialogic in nature.

*Univocal* discourse allows an individual to accurately transmit information to another person (Lotman, 2000). This pattern is characterized by teachers telling students the content to be learned, showing them the way to solve problems, painting a picture of mathematics that seems to reflect the belief that mathematics is inert, structured, and immutable. In contrast, *dialogic* discourse incorporates the belief that mathematics is socially constructed and negotiated, and, as such, creates new interpretations as individuals negotiate meaning (Wertsch, 1998). We theorize that how a teacher solicits student participation and determines acceptable responses influences the *function* of the discourse; i.e., whether it is predominately univocal or dialogic. These solicitations and judgments reveal teachers' goals and their deeply-held beliefs. In addition to the function of discourse, teachers also choose to use particular forms of talk during instruction.

Truxaw (2009) identified four forms of talk within a whole-group discussion setting. They include:

1. monologic talk in which the teacher does all of the talking without expecting any responses;
2. leading talk in which the teacher controls the discussion and the direction of the discourse;
3. exploratory talk in which the teacher poses questions without anticipating the direction of subsequent discourse; and
4. accountable talk in which the teacher pushes for appropriate use of mathematical language, generalizations and justifications.

Here, we describe the teacher and student discourse we observed in four pre-calculus classes. Following this brief discussion, we will focus our attention on two of the four classes where we observed a similar triadic discourse pattern. However, we realized that the effects and results in these two classrooms were startlingly different, particularly in the building of a community of learners (COL). We conjecture that the main reason for this difference was in the beliefs that the two instructors held about the role of the instructor, the role and responsibilities of the student, and the form and function of the classroom discourse. These combined to impact the learning environment.

## METHODS

This qualitative study is part of a larger research project that investigated how classroom structure and teacher discourse in four sections of a pre-calculus course supported student learning. For the project, both quantitative and qualitative data were gathered. Analysis of students' test scores on four common examinations showed that students in the PBL section outperformed the students in the other sections (see Cooper & Olson, 2008).

The purpose of this study was initially to compare and contrast the different instructional strategies to determine that led to higher student engagement and achievement. A case-study design (Merriam, 1998) with cross-case analysis (Miles & Huberman, 1994) enabled us to characterize the instructors' pedagogy and to contrast these pedagogies across cases. Data were collected from five sources during the fall semester. These sources included:

1. weekly reflections;
2. monthly video recording on two consecutive days;
3. field notes made by the videographer;
4. informal interviews; and
5. semi-structured follow-up interview.

Constant comparative methods (Merriam, 1998) were used to code, analyze, and collapse the data to identify emergent patterns of discourse (Truxaw, 2009). These patterns were then displayed in a conceptual matrix with illustrative examples in each cell (Miles & Huberman, 1994). Initially, our analysis focused on interpreting the discourse captured on the videotapes and triangulating our interpretations with the instructors' reflections, videographers' field notes, and interviews.

## CONTEXT

This study took place in a large university in the United States serving 18,000 students. A placement examination identified entering students who needed prerequisite skills to be successful in calculus. The students were randomly assigned to a section of pre-calculus. The student composition (scores on the placement exam, gender, career aspirations) was consistent across sections. Quantitative analysis indicated that students in the PBL section outperformed the students in the other sections and this disparity increased over time (Cooper & Olson, 2008).

The four pre-calculus sections that were the focus of this study were initially constrained by class sizes of 70, classrooms with auditorium seating, and scheduling that required them to meet four days a week in 50-minute classes over a 16-week semester. With departmental funding we were able to modify three of these sections while the fourth section (typical lecture) remained unchanged. Four instructors volunteered to participate in the project and are referred to by pseudonyms for each instructor using the first letter of the approach for reference. The instructor for the typical lecture section is Trudy, for the small class size section is Sam, for the SCG section is Sara, and for the PBL section is Pam. The instructors were respected by students and had strong evaluation ratings in pre-calculus during the previous semester.

## RESULTS

This study examined the interactions between the instructors and their students to characterize:

1.  the function and forms of their discourse;
2.  the use of questions;
3.  the instructor's expectations of students' success and responsibilities; and
4.  the role of students in class.

We summarize the key features of the instructional practices in each class in Table 1. We describe the pedagogical practices of each instructor, and then contrast the practices of two instructors to illustrate pedagogical practices that may increase student achievement. Thick description is used to portray the role of discursive practices that fostered this development.

## ANALYSIS

We theorize that how and whether and to what extent a teacher solicits student participation and determines acceptable responses influences the function of the discourse. These solicitations and judgments reveal whether a teacher's enacted beliefs are consistent with his or her stated beliefs. The intersection of the teacher's questions and judgments with his or her beliefs determines the function of the discourse, defined as univocal or dialogic.

**TABLE 1. Characterization of Instructors' Practices during Intervention (Fall 2007)**

| | Christy (Typical) Trudy | Matt (Small Class) Sam | Liz (SCG) Sara | Lisa (PBL) Pam |
|---|---|---|---|---|
| Functions of discourse | Univocal | Univocal | Univocal | Dialogic |
| Forms of discourse and exchange | Monologic Leading Triadic exchanges | Monologic Few triadic exchanges | Monologic Leading Few triadic exchanges | Leading Exploratory Triadic exchanges |
| Questions | Leading | Did not ask them | Leading | Leading Affective Few reflective |
| Expectation | Half of the students will be successful | Only some students will learn | With enough work, every student can learn | Every student can learn |
| Responsibility to learn | Instructor shows students what to do | Instructor shows students what to do | Instructor is responsible for student success | Instructor and students share the responsibility |
| Student role | Engaged but not autonomous students who are anonymous and invariant from year-to-year | Passive observers who listened and watched the instructor solve problems | Individuals who worked alone, following the instructor's model | Active, attentive participants who engaged in note-taking and problem-solving |

What follows is a description of each class.

## Trudy's Traditional Lecture

Trudy believed her responsibility as an instructor was to "present the material in as clear a manner as I can, require homework, and hold them to the syllabus." She demonstrated how to solve problems and students were responsible for learning the material by completing homework. For Trudy, the function of discourse was to support student learning by telling students what they needed to learn and explaining the steps clearly. Thus, her discourse was univocal, and her desire to make things clear and accurate translated into her use of monologic talk with leading questions.

Trudy stood in front of an overhead projector and white board. She talked with a loud, confident voice that communicated both her knowledge of the content and experience teaching the course over several semesters. The discourse was a constant flow of triadic exchanges in which several students usually responded. Occasionally students were silent and Trudy would either rhetorically answered her question or she rephrased it to something that they could answer. At one point during a class, Trudy wanted to correct a common mistake when solving for a variable, $5 = a(0 - 2)^2 - 3$ and asked, "What would happen if we squared it first?"—silence—"Could we square each of them individually?" One student softly responded, "No." She wrote on the board, $5^2 = [a(0 - 2)^2]^2 - 3^2$, and replied, "No. So you can't just say five squared, this squared $[a(0 - 2)^2]$ and this $[3^2]$." Trudy recognized an opportunity to correct a misconception and used monologic talk to tell students that squaring the individual components of the equation was incorrect. She chose to do this rather than to engage in exploratory talk that would allow the students to reconstruct the process of squaring. After her statement, a student began to ask, "What if..." and was interrupted by Trudy. Perhaps she did not hear the student or perhaps the question broke the triadic flow in which the instructor asks the questions.

Trudy posted problems on the overhead projector and engaged students in the problem-solving process by asking them, "What do you want to do?" or "What do we know?" These questions had the potential for exploratory talk that would allow students to guide her through the problem solving process. However, she ignored responses that did not agree with her anticipated strategy. By silencing students' diverse ideas, she eliminated the possibility of developing a discourse with a dialogic function in which students had a role in creating the solution.

Trudy asked leading questions that were used to guide students through an efficient solution strategy. Correct responses were phrases or single words and she affirmed it by extending the phrase into a sentence. For example, she drew a triangle on the board for the problem, Find cot(arcos $6/x$). She asked, "What sides do we know?" A student responded, "Adjacent and hypotenuse." She rephrased and extended his response to, "So we know that cosine is adjacent over hypotenuse so the adjacent is 6 and hypotenuse is $x$." Some of the leading questions were well rehearsed and triggered by the word *normal*. When she asked a question like, "What do we normally do next?" students responded in chorus, "Rationalize," "Plug in," or "Pythagorean theorem."

The lecture was a continuous flow of both leading questions and question fragments with pre-determined responses that kept students engaged in class, but did not provide them with opportunities to independently engage in problem solving. At one point a student questioned her statement that $y^2 = x^2 - 36$ can not be rationalized. Rather than listening to the student, she immediately cut him off stating, "You can't simplify" and continued solving the problem. Alternative ideas were not tolerated due to "time limitations." Trudy moved on following an itinerary with no time for unexpected detours. Thus, deviations that might help students understand the process or correct misconceptions were not allowed. There were no improvisational moves partly because no time was allotted for them. There seemed to be no need for improvisation in a class in which the instructor controlled the discourse and students listened, took notes, and responded on cue.

### Sam's Small Lecture

Sam reviewed the daily content and presented examples. He showed students how each example was solved pointing out important mathematical steps and choices made. Sam asked the class questions that drew their attention to an algebraic principle and students responded in either a chorus or a single voice called out a response. Students were assigned homework problems that required them to practice finding solutions to problems that were similar to the ones presented in class.

Sam's class was conducted with him standing in the front of the class demonstrating on graph paper at the overhead how the different problems might be solved. He talked in a monotone voice and it was difficult to hear what he said. His discourse was not modulated nor did his voice carry any inflection. Nothing in his univalent, univocal discourse stood out. He made little attempt to engage the students in the class although they seemed to try to follow along and occasionally offer next steps to his leading questions. When he asked a question, he immediately filled in the answer himself. He demonstrated problem after problem, pointing out the highlights of each, noting the places in each one where care needed to be exercised, stating what he expected on the test and the form in which answers should be presented. This univocal discourse provided compelling evidence that Sam's belief about the role of the teacher is to convey information and demonstrate skills. He expected students to make sense of the mathematics at a later time, outside the classroom.

One of Sam's lectures was on transformations of graphs—given a complicated function how could it be graphed by recognizing it as the transformation of a simpler function whose graph was already known. He demonstrated example after example to ensure that his students had seen every possible type of transformation, and were clear on how to apply the stated rules. Sam did not talk about a general approach to this type of problem; each example was unique. Thus, students were left to learn a set of unconnected procedures, without the benefit of the underlying unifying structure. He did not ask anything of them during class, students were passive. Most students said nothing; a few ventured an occasional answer. Sam exemplified the sage while students acted as the audience observing his performance. Each performance was predictable and consisted of repetitive exercises in rule-following.

## Sara's Small Group Collaboration

The SGC class met for three of the days in a traditional lecture hall and the fourth day in a classroom where two to four students might interact at tables. The time spent in the lecture hall was used by the instructor to:

1. explain new material;
2. solicit prior understandings; and
3. work through examples illustrating the concepts and factual knowledge, using a traditional lecture format.

On the fourth day, during SGC, Sara distributed problems and students began to work on them individually. Their tables became desks and students occasionally collaborated with each other. Some students asked a neighbor for help or to check their answers. The instructor did not interact with many student groups as it was difficult to circulate around the packed classroom. Sara hoped that a community of learners would evolve and that students would work together outside of class; however there was no deliberate structure in place to effect this outcome. In actuality, few students met outside of class.

Sara's teaching style was such that she rarely called upon individual students to answer her questions, but instead accepted and encouraged volunteer responses to general questions. She frequently threw questions out during her lecture, and this had a tendency to keep students engaged in the problem solving. One of Sara's lectures was devoted to setting up and using models of exponential growth and decay. She opened the class with an explanation of modeling problems in general, in the hope of motivating the students to want to learn more about modeling and to understand why they were learning about modeling.

She recalled for the students the function $f(t) = e^t$ and explored their understanding by asking them to describe the graph, sketch it and tell her its important features. Several students responded to her leading questions. She clearly had a definite goal and purpose to her questioning, and had particular answers in mind. After summarizing the students' responses, sketching the graph and labeling important features, she defined the exponential growth model $N(t) = N_0 e^{kt}$.

She then asked several leading questions concerning the roles and meanings of $N_0$, $k$, and $t$, and posed the question, "What would that give us if we plugged in a value of $t$ to $N(t)$?" Several students responded that this would give the value of N at time $t$. She proceeded to write a problem on the board. After her leading questions about the model, she was then in a position to use exploratory discourse to enable students to guide her towards a solution. Instead, she used monologic (telling) discourse: "The way we do this is…", thus shutting out any inventiveness or creativity on the part of the students in determining a solution strategy. In the course of her monologic talk she asked small leading questions such as "What is $N_0$?"; "Do we know $k$?"; "What is $t$ measured in?" These questions had the effect of keeping most students engaged in the process. However, it did not give them practice solving these types of problems independently.

At another point during the lecture, she asked a question that seemed out of character: "Why do we have to do that?" (in reference to converting a time of three hours to minutes). At another instant she finished a problem and asked if anyone knew a different way to figure it out. This appeared on the surface to be an invitation to accountable talk allowing students to be fully engaged and share their knowledge. However, when an alternate strategy was offered by a student, Sara did not accept it. It became clear that she had a preconceived alternate method in mind that she wanted to show the class.

In problem after problem, she led students step-by-step but never let them off the "lead rein" to work on their own. Observing the lessons, one had the feeling of a very closely controlled and choreographed performance. The majority of her talk throughout the lesson was best characterized as telling or monologic talk, or, in the case of questions, leading questions. Although students seemed to be quite comfortable jumping in with an answer to her questions, she seldom gave them very much time to think about the answer, and more often than not she treated her questions rhetorically and answered them herself.

## Pam's Problem-Based Learning

Pam believed that every student can learn and her role was to provide them a framework from which they could study and learn. She remarked, "The biggest problem is that students don't know how to study and take notes... I help them by modeling the notes that will help and teach them study skills." During the lecture periods, Pam introduced a new topic or section by posing a question which required students to apply previous skill or knowledge in a new situation. The students explored the new content for a few minutes independently and then the instructor asked several students to share what they found.

For example, when Pam introduced trigonometric functions, she posed the following problem, "Are trig functions distributive? In other words, what would $\sin(\pi/3 + \pi/6)$ equal if it was distributive?" Several students responded, "$\sin \pi/3 + \sin \pi/6$?" Then Pam replied, "So, what I want you to do is figure out whether $\sin(\pi/3 + \pi/6) = \sin \pi/3 + \sin \pi/6$." Students worked on the problem for a minute before they indicated that they were ready for a discussion on the problem by giving her a thumbs-up signal. The talk became exploratory as she solicited different approaches by asking, "Does anybody have a general way to show this concretely?" or "What do you think?" After one student provided a suggestion, she responded, "Yeah, you could... Anyone else?" encouraging several students to share ideas. Then, she selected one strategy and reaffirmed other methods, "There are lots of ways to approach this kind of problem, this is just one way." Thus, Pam immediately engaged students in an investigation in which they helped her explore a mathematical idea.

After Pam selected a strategy, she asked leading questions, "What is...?"; "What do we get?"; "Which reduces to...?" or "What does that tell us?" These questions guided students through the steps to solve the problem and arrive at a conclusion. Rather than seize the opportunity for accountable talk in which students summarized the solution, Pam stated the findings and provided a written generalization on the board. Students copied the notes into their notebooks and occasionally asked a question to clarify a point. Pam responded to each question without asking the students to address a peer's question. In this way, she maintained control of the discourse and deviations from her plans were minimized.

Pam frequently asked students whether they wanted to solve problems independently "Do you want to go at it alone?" or if they wanted to work through it together. Usually, the class indicated that they wanted to work on it alone. Students began independently and talked to a neighbor to check an answer or for help. They were encouraged to be creative and engage in problem solving as partners rather than passive observers. Pam used students' responses help her decide what strategy to emphasize or to clarify an idea. Interactions between Pam and her students were like conversations that one might have over a cup of coffee with a tutor. Pam relied on leading questions to help students work through the correct algebraic procedures for simplifying an expression or solving a problem. Throughout the class, the talk was collaborative and supportive.

## COMPARING TRUDY AND PAM

Analysis of students' exam scores in the larger project revealed that scores were significantly and consistently higher in the PBL section throughout the semester (Cooper & Olson, 2009). Cross-case analysis of the four instructors' pedagogies revealed two characteristics that were unique to the PBL section: a notable sense of community and higher student achievement. Pam's class had statistically higher test scores and we noticed that her classroom environment was different from the other three sections. Analysis of the three sections had indicated that Trudy's discourse pattern was quite similar to Pam's. Both Trudy and Pam employed an initiation, response, evaluation (IRE) discourse pattern. They both used leading questions to control student responses and ensure that unexpected answers and approaches were not considered. In addition, both instructors provided generalizations for students in their summaries, obviating the need for students to synthesize material for themselves. These similarities are evident in the preceding discussion of each instructor's practice. For these reasons, we chose to focus more closely on two particular classrooms: Pam's and Trudy's.

So what could explain the difference in student achievement? This section focuses on three key components of teaching that influenced student learning in these two classrooms: *teacher beliefs* about their roles and that of their students; the *classroom discourse*; and the *classroom environment*. Analysis of these components led us to create the model in Figure 1.

It struck us immediately that Trudy's and Pam's beliefs about the role of teachers and students were fundamentally different. Trudy believed that it was the teacher's role to prepare students for the test in an orderly and logical fashion. She expected her students to attend to the lecture and take appropriate notes. Also, based on her past experience teaching the course, she believed that about half of the students would succeed in passing the course. Trudy believed that a teacher had little influence on the pass rate as long as he or she presented the material coherently. The responsibility for

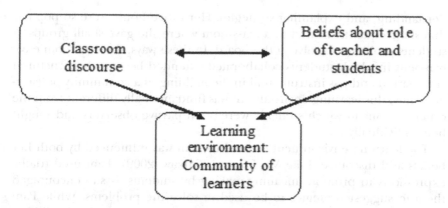

**FIGURE 1: The interactions among discourse, beliefs and learning environment.**

learning resided solely in the students' effort; the ones who put in the effort were the ones who would pass. Thus, the function of Trudy's discourse was monologic; it served to convey the important information. These beliefs and the monologic function of her discourse created a learning environment in which students listened, watched, and engaged in taking notes that would enable them to solve similar problems later. During office hours, when Trudy's students asked for help, she told them which strategy to use and led them through the solution process. Trudy never called on individual students in class. It was her belief (personal interview) that this would be too intimidating for the student.

In contrast, Pam believed that a teacher does make a difference, both in making the content understandable to students and working with them in a collaborative process, and in motivating them to succeed. She used inclusive language that invited students to participate and encouraged them to be successful by constantly affirming their suggestions. She often used the words "we" and "let's" as she worked through a problem solution with her students. The use of these words communicated clearly her belief that they were all working together and success was determined by their joint effort. These beliefs influenced her discourse to be dialogic in function, characterized by its conversational nature. The role of students in Pam's class was to actively solve problems during the frequent breaks in the lecture and to engage in a conversation at her invitation.

Pam carefully drafted class notes on the board and expected students to copy them down. Students appreciated these notes that synthesized the information into coherent summaries from which they could study. Students in Pam's class gathered twenty minutes before class to work on problems. Pam met with these small groups answering questions, checking for understanding, and explaining strategies. Her office hours were so popular that she frequently moved to a classroom where she gave small groups of students problems to solve at the board. In these ways, the classroom environment in which students collaborated extended beyond the structured class period and was instrumental in the building of a community of learners. Thus, this learning environment was fundamentally different from the other sections in which students were often passive observers and sought help individually.

The learning environment created by Pam was influenced by both her beliefs and discourse. Like Mr. Larson (Truxaw, 2009), Pam used triadic expressions to promote meaning-making by students as she encouraged them to suggest strategies to be used to solve the problems. While Pam routinely and consistently engaged students in exploratory talk, she did not invite them to share in accountable talk. She assumed the responsibility to synthesize and summarize the important mathematical ideas discussed in the lesson, rather than encouraging them to synthesize and generalize for themselves as a way of sense-making. Pam's class may have been all the more powerful and successful had she been able to engage her students in this way. However, we believe that what made the difference in Pam's class was the way that her beliefs, the discourse, and the environment that pervaded the classroom, acted in concert to establish a community of learners (COL).

According to Sherin, Mendez, and Louis (2004) fostering a COL in a mathematics classroom requires three things. First, the teacher must move away from the concept of mathematics as a combination of processes and facts. Second, the role of the teacher in the classroom must be reconceptualized by instructors and they must begin to see their role as going beyond the mere implementation of reform-based curricula like problem-based learning (PBL) to engaging in activities that fosters dialogic discourse, justification of mathematical truths and generalizations of mathematical ideas. Third, the teacher must recognize that the core principle in fostering a COL is creating a discourse community in the classroom. In short, a discourse community requires students to express their ideas, listen to and respect the ideas of others, and to evaluate such ideas and solutions in meaningful ways (Sherin, Mendez, & Louis, 2004).

In conclusion, although the discourse patterns of these two teachers were quite similar, the interplay between the classroom discourse and the teachers' beliefs about teachers' and students' roles impacted the learning environment. Pam created a learning environment that *encouraged* students to express their ideas and listen to and respect the ideas of others. With appropriate support and mentoring, Pam could have developed this into a COL. However, she did not establish a COL because students had the option to participate in the classroom discourse and they did not evaluate each others' ideas as they collectively built mathematical meaning. Further research is needed to investigate how the development of a COL can be supported in post-secondary classrooms and whether student achievement increases when instructors establish a COL.

## REFERENCES

Biddle, B. J., & Berliner, D. C. (2002). *What research says about small classes and their effects*. Retrieved April 20, 2008 from http://www.wested.org/online_pubs/small_classes.pd

**Boaler, J. (1998).** Open and closed mathematics: Student experiences and understandings. *Journal for Research on Mathematics Education, 29*(1), 41–62.

Boaler, J. (2000). *Multiple perspectives on mathematics teaching and learning*. Norwood, NJ: Abex.

Cazden, C. B. (2001). *Classroom discourse: The language of teaching and learning*. New Portsmouth, NH: Heinemann.

Cobb, P. (2000). From representations to symbolizing: introductory comments on semiotics and mathematical learning. In Cobb, P., Yackel, E., & McClain, K. (Eds.) *Symbolizing and communicating in mathematics classrooms* (pp. 17–36). Hillside NJ: Lawrence Erlbaum.

Cohen, E. (2006). Science and engineering indicators 2004. *Higher Education in Science and Engineering*. Retrieved April 8, 2008 from http://www.nsf.gov/statistics/seind04/c2/c2s2.htm.

Cooper, S., & Olson, J. (2008). *Math solutions: The impact of classroom structure on teaching and learning in pre-calculus classes*. Report. Washington State University.

Davis, B., & Simmt, E. (2003). Understanding learning systems: Mathematics education and complexity science. *Journal for Research in Mathematics Education, 34*(2), 137–167.

Erickson, D. K. (1999). A problem-based approach to mathematics instruction. *Mathematics Teacher, 92*(6), 516–521.

Glass, G. V., Cahen, L., Smith, M. L., & Filby, N. (1982). *School class size: Research and policy*. Beverly Hills, CA: Sage.

Grouws, D. A., & Cebulla, K. J. (2000). *Improving student achievement in mathematics* (Report No. SE 064 318). Columbus, OH: ERIC Clearinghouse for Science, Mathematics, and Environmental Education.

Hiebert, J., Carpenter, T. P., Fennema, E., Fuson, K., Human, P., Murray, H., Olivier, A., & Wearne, D. (1996). Problem solving as a basis for reform in curriculum and instruction: The Case of Mathematics. *Educational Researcher, 25*(4), 12–21.

Hiebert, J. Carpenter, T. P., Fennema, E., Fuson, K., Human, P., Murray, H., Olivier, A., & Wearne, D. (1997). Making mathematics problematic: A rejoinder to Prawat and Smith. *Educational Researcher, 26*(2), 24–26.

Higher Education in Science and Engineering. (2004). *Science and engineering indicators 2004*. Retrieved April 8, 2008, from http://www.nsf.gov/statistics/seind04/c2/c2s2.htm.

Jarvis, T. (2007). Class size and teacher effects on student achievement and dropout rates in university-level calculus. Retrieved April 7, 2008 from http://www.math.byu.edu/~jarvis/class-size/class-size-preprint.pdf.

Kazemi, E., & Stipek, D. (2001). Promoting conceptual thinking in four upper-elementary mathematics classrooms. *Elementary School Journal, 102*(1), 59–80.

Kenny, G., Kenny, D., & Dumont, R. (1995). *Mission and place: Strengthening learning and community through campus design.* Westport, CT: Greenwood.

Krulik, S., & Rudnick, J. A. (1999). Innovative tasks to improve critical- and creative-thinking skills. In Stiff, I. V. (Ed.), *Developing mathematical reasoning in grades K–12* (pp. 138–145). Reston. VA: National Council of Teachers of Mathematics.

Kyeong, R. (2003). *Problem-based learning in mathematics.* ERIC Clearinghouse for Science Mathematics and Environmental Education. Retrieved April 15, 2007 from http://www.ericdigests.org/2004-3/math.html.

Ladson-Billings, G. J. (1995). Toward a theory of culturally relevant pedagogy. *American Education Research Journal, 35,* 465–491.

Lave, J. (1996) Teaching, as learning, in practice. *Mind, Culture, and Activity, 3*(3), 149–164.

Lewellen, H., & Mikusa, M. G. (February 1999). Now here is that authority on mathematics reform, Dr. Constructivist! *The Mathematics Teacher, 92*(2), 158–163.

Lotman, Y. M. (2000). *Universe of the mind: A semiotic theory of culture* (A. Shukman, Trans.). London: I. B. Tauris & Co Ltd. (Original work published in 1990).

Lutzer, D. J., Rodi, S. B., Kirkman, E., & Maxwell, J. W. (2007). *Statistical abstract of undergraduate programs in the mathematical sciences in the United States.* Providence, RI: American Mathematical Society.

Merriam, S. B. (1998). *Qualitative research and case study applications in education.* San Francisco, CA: Jossey-Bass.

Miles, M. B., & Huberman, A. M. (1994). *An expanded source book: Qualitative data analysis* (2nd ed.). Beverly Hills, CA: Sage.

Rasmussen, C., Kwon, O. N., Allen, K., Marrongelle, K., & Burtch, M. (2006). Capitalizing on advances in mathematics and K–12 mathematics education in undergraduate mathematics: An inquiry-oriented approach to differential equations. *Asia Pacific Educational Review 7*(1), 85–93.

Sfard, A. (2008). *Thinking as communicating: Human development, the growth of discourses, and mathematizing.* New York: Cambridge University Press.

Sherin, M. G., Mendez, E. P., & Louis, D. A. (2004). A discipline apart: the challenges of 'fostering a community of learners' in a mathematics classroom. *Journal of Curriculum Studies, 36*(2), 207–232.

Shoenfeld, A. (1998). Reflections on a course in mathematical problem solving. *CBMS Issues in Mathematics Education, 7,* 81–102.

Truxaw, M. (2009). Orchestrating whole group discourse to mediate mathematical meaning (Chapter 7 in this monograph). *The Montana Mathematics Enthusiast.* monograph #9. Charlotte, NC: Information Age Publishing.

Turner, J., Meyer, D., Anderman, E., Midgley, C., Gheen, M., & Kang, Y. (2002). The classroom environment and students' reports of avoidance strategies in mathematics: A multimethod study. *Journal of Educational Psychology, 94*(1), 88–106.

Walkerdine, V. (1997). Redefining the subject in situated cognition theory. In Kirshner, D. & Whitson, J.A. (Eds.) *Situated cognition: Social, semiotic, and psychological perspectives* (pp. 57–70). Mahwah, NJ: Lawrence Erlbaum.

Wertsch, J. V. (1998). *Mind as action.* New York: Oxford University Press.

## CHAPTER 4

# "YEAH, BUT WHAT IF...?"

## A Study of Mathematical Discourse in a Third-Grade Classroom

### Karen M. Higgins, Cary Cermak-Rudolf, and Barbara Blanke

This case study[1] is about Cary and her third-grade classroom, the site of a discourse-rich community of young learners. The data were collected over three years and came from the following sources: notes from observations and site visits to Cary's classroom and school, formal and informal interviews with Cary and her students, journal entries, personal reflections, artifacts from student work, and transcripts of classroom episodes.

The researchers found the following dispositions and experiences important to understanding discourse and implementing it purposefully in the mathematics classroom: a desire to grow as a teacher and learner of mathematics; the ability to collaborate with and observe others, using reflective protocols that focus on student thinking; a willingness to decentralize practice; and experiences with high-level discourse as a mathematics student.

Building sociomathematical norms for robust mathematical discourse happened through viewing disequilibrium as learning opportunities, structuring group work with attention to minimizing status, pressing for justification and generalization around the lesson's core mathematics, and stressing authentic questions from the teacher and students.

[1]The Oregon Mathematics Leadership Institute Partnership Project is funded by the National Science Foundation's Math Science Partnership Program (NSF-MSP award #0412553) and through the Oregon Department of Education's MSP program.

The researchers believe there are implications from this research for the teaching and learning of mathematics that are important in all teacher training programs and professional development.

**KEYWORDS:** discourse; sociomathematical norms; mathematics teaching and learning

As you walk into the middle of Cary's mathematics lesson, her third-grade students are nestled around a task on equivalent fractions. You hear comments from students: "Can you prove it?" "Are you sure it works that way?" "I don't see it. Can you show me again?" At a pivotal moment during these group discussions, Cary pulls students together for a plenary where she has carefully selected students for presenting their solution strategies and accompanying rationalizations. What follows are bursts of discourse amongst the students that culminate into mathematical understandings that leave the students with a feeling of satisfaction, joy, and a growing desire to learn more.

A visit to such a classroom leaves the observer awed and dumbfounded. One walks out with many questions about how a teacher even begins this journey with her students. What did it take to get third-grade students to talk about their mathematical thinking at this level? What was Cary's role as a teacher? How did she start? How did her students start? What worked? What didn't? Do the children even have a clue as to the mathematicians they have become?

## BACKGROUND

For the past four years, Cary was involved in a NSF-funded Mathematics and Science Partnership (MSP) grant which brought together over 180 K–12 teachers, 90 administrators, and 40 higher-education faculty. The vision of the project was to provide support for systematic mathematics reform. The project provided three-week residential summer institutes (2005, 2006, 2007) that combined rigorous mathematics content coursework with collegial leadership development. The fourth summer involved a three-day symposium that highlighted the impact of the grant on the schools involved with the project.

In year one, the instructional focus of the collegial leadership course work was on the parallel between the work in teachers' classrooms and their mathematics professional learning communities. The focus of years two and three was on their continued development as teacher leaders and facilitators of mathematics professional development. The mathematics strand of the grant involved graduate-level course work in mathematics. Six areas were covered over the course of the three summers: Discrete Mathematics, Algebra and Functions, Number and Operations, Data and Chance, Measurement and Change, and Comparing Geometries.

In addition to the summer institutes, each school in the grant was funded for four half-day site visits over the course of each year. These visits were to support and push the teachers as they began changing their teaching pedagogy and attempted to create discourse-rich teaching and learning environments for their students. The third year of the grant focused on the teachers' professional development work within their schools.

A major focus of the grant was to increase the quality and quantity of mathematical discourse in the mathematics classrooms of the schools involved in the grant. It was the belief that by increasing student discourse and allowing students the opportunity to talk about their thinking, the vision of the grant would be realized:

a.  Increase mathematics achievement of all students in participating schools;

b.  Close achievement gaps for underrepresented groups of students;

c.  Increase enrollment and success in challenging mathematics course work that support state and national standards through coherent, evidence-based programs.

This study is a story about Cary and her third-grade classroom over a three year span. In particular, it investigated the following three questions:

1.  How does a teacher begin to learn about using discourse purposefully to improve mathematics teaching and learning?

2.  How do norms of mathematical discourse evolve and how can teachers influence the development of these norms? and

3.  What does a discourse-rich classroom look like and how is it different from more traditional classrooms?

One of the authors of this paper (Higgins) was involved in the site visits at Cary's school and was one of her instructors for the leadership course at the summer institute.

The data for this study were collected over three years. The focus on discourse influenced the selectivity of what was collected and how. Overall, data came from the following main sources and field texts:

- notes from observations and site visits to Cary's classroom and school;
- formal and informal interviews with Cary and her students, journal entries (researchers and students);
- personal reflections (researchers);

- artifacts from student work; and
- transcripts of classroom episodes.

The transcripts were collected as part of Blanke's dissertation research in Cary's classroom. The three authors viewed themselves as "in the middle of a nested set of stories" (Clandinin & Connelly, 2000, p. 63) as they worked together and supported each other through this journey of reflection, research, and writing, unfolding the story of an amazing teacher and her third-grade children who knew how to "talk" mathematics.

## MATHEMATICAL DISCOURSE

Mathematical discourse is an iterative and interactive process in which students engage in conversations about mathematical ideas at various cognitive levels. According to Ball (1991):

> Discourse is used to highlight the ways in which knowledge is constructed and exchanged in classrooms. Who talks? About what? In what ways? What do people write down and why? What questions are important? Whose ideas and ways of knowing are accepted and whose are not? What makes an answer right or an idea true? What kinds of evidence are encouraged or accepted? (p.44)

The term discourse is used in many contexts and broadly understood it is considered the study of actual language usage in specific communicative contexts (Schwandt, 2001). Discourse analysis differs from the idea of conversation analysis as it strictly focuses on the content of the talk rather than the linguistic structure. Discourses can also be described as practices composed of ideas, ideologies, attitudes, courses of action and terms of reference that systematically constitute the subjects and objects of which they speak (Foucault, 1972). Foucault's assertion that discourse shapes subjects and their worlds supports a possible link between discourse communities and the paradigm shifts that teachers choose and how these shifts affect their perceptions of self-efficacy as teachers of mathematics.

Discourse focuses on the act of articulating mathematical ideas or procedures through talking, asking questions, and writing. These articulations can occur in different configurations within the classroom (NCTM, 2000):

- Student to Teacher: The student primarily addresses the teacher
- Student to Student: The student addresses another student
- Student to Class/Group: The student addresses the entire class or a small group of students
- Individual Reflection: The student documents his or her reflections about mathematics, generally in writing.

Student and teacher questions lead to explanations and justifications that may be challenged and subsequently defended within discourse communities. This process may lead to the formation of new generalizations or conjectures that initiate new occurrences of different types of mathematical discourse.

Teachers play a crucial role in shaping the discourse in their classrooms through the signals they send to their students about what is valued about mathematical knowledge as well as ways of thinking and knowing about mathematics (Ball, 1991). Creating and maintaining discourse environments is a complex endeavor for teachers (Sherin, 2002). It is not enough to encourage students to discuss their ideas and converse with each other; teachers also must ensure that these discussions are mathematically productive. Research has shown that asking teachers to use discourse requires teachers to develop a new sense of what it means to teach mathematics and of what it means to be an effective and successful mathematics teacher (Sherin, 2002; Smith 1996). This supports the need to better understand the teacher's role in choosing to change and effectively implement math reform, specifically mathematical discourse.

### Why Discourse?

Kazemi claimed (1998) when teachers helped students build on their thinking, student achievement in problem solving and conceptual understanding increased. The National Council of Teachers of Mathematics Principles and Standards (NCTM, 2000) support improved teaching in mathematics and visualize classrooms where teachers frequently engage in mathematical discourse with their students, delivering fewer monologues as an integral part of effective teaching of mathematics (Springer & Dick, 2006).

According to Stein (2001), one of the most difficult recommendations of NCTM's Standards is to orchestrate mathematical discourse—moving from teacher-centered classrooms to ones centered on student thinking and reasoning. Most teachers have not experienced learning mathematics in this way. So why do teachers choose this paradigm shift when teaching mathematics? How do they become teachers who encourage discourse and facilitate communities of learners?

Many studies have shown that what constitutes good mathematics teaching is consistently controversial and continues to be controversial in the mathematical world (Ball & Bass, 2003; Franke, Kazemi, & Battey, 2007; Hiebert, Morris, & Glass, 2003; Smith, 2002). Describing the state of the K–12 mathematical-teaching experience and the knowledge needed for effectively teaching mathematics is complex to say the least (Ball & Bass, 2003)

## DISCOURSE AND CARY

VanZoest and Enyart (1998) claimed that discourse could be a problem area for teachers when they did not realize how important it was or when they had not seen or experienced it themselves. But, even though teachers were convinced of its importance and they had a desire to change their classroom practice, the biggest obstacle was a lack of awareness of how to make meaningful discourse a reality in their classrooms.

Although there were readings related to discourse in Cary's leadership courses, she claimed it was her first OMLI mathematics course, Discrete Mathematics, which had the greatest impact on her understanding of discourse:

> I HATED it my first year of the institute. It made me feel inadequate, and I was. I was on many levels. Discourse makes you think. It is different than being force-fed information. It allows you to build and scaffold knowledge at your level. It pushes you to make sense of it all. You have to have a deep understanding that then, hopefully, gets pushed even further. My discrete instructor GOT IT! After I left that class, I knew what it looked and felt like. Watching our instructors struggle to ask the best question allowed me not to take it so personally, but to see me as a work in progress. (C. Cermak-Rudolf, personal reflection)

Cary had the opportunity to sit down and plan a lesson with another third-grade teacher from her district who was also involved in the grant but in another school. After they planned the lesson using the "Lesson Planning Framework" presented at OMLI, Cary co-taught the lesson with her in her colleague's classroom. After this teaching experience, they reflected on their questioning strategies and student outcomes, and fine-tuned the lesson for the second co-teaching experience in Cary's classroom. In her summer reflection after the first year, Cary said this about the experience:

> The ability to be in another person's classroom and absorbing their use of discourse and math-questioning strategies was an amazing event....This helped me on many levels. I was able to see and learn new questioning strategies. I was able to observe a question being asked and see the students' cognitive response. (C. Cermak-Rudolf, personal reflection)

The site visits also impacted Cary's understanding of using discourse purposefully. During the first two years, Cary claimed Higgins knew how to push her students and herself to the next level. "She would observe 'things' that I hadn't seen, or pose a question for me to see 'it' in another way. It could be a student thought, my questioning strategies, or simply my body language."

Many of the site visits focused on Cary and her classroom. These site visits often involved the principal in Cary's school. The pre-observation conferences prior to Cary's teaching utilized OMLI's "Lesson Planning Framework." This included doing the mathematics of the lesson together as well as focusing on various aspects of the lesson, such as the core mathematical goals, ways to enhance the cognitive demand of the tasks, the anticipation of students' responses to the lesson, and potential misconceptions. The "Student Discourse Observation Protocols" were used as part of the observation process, capturing as much student discourse as possible during a lesson. It is important to note that the focus was on students and their thinking, not teacher moves or teacher talk. In order to have an understanding of the level and trends of student discourse happening in Cary's classroom, the observers and Cary sat down together after the observation to review the data and categorize the statements into one of three areas:

1. procedures/facts;
2. justification; or
3. generalization.

Quite often, this categorization process led to powerful discussions of what constituted rich discourse in the classroom. Whatever the data indicated, the observation team was able to ask questions of each other that strengthened their understanding of the levels of discourse. They began to see discourse as a continuum rather than three discrete categories. Even though the focus at the beginning of the project was on student thinking, as their trust level increased they began looking at the teacher moves that elicited or hindered productive discourse in Cary's classroom. Their collective co-inquiry questions honed them in on certain issues that provided all of them with a deeper understanding of the teaching and learning of mathematics and furthered their collective understandings of the depth of the sociomathematical norms present in Cary's classroom.

## ESTABLISHING NORMS OF MATHEMATICAL DISCOURSE IN CARY'S CLASSROOM

According to Rasmussen, Yackel, and King (2003), social norms refer to those aspects of the classroom social interactions that become routine. "Every class, from the most traditional to the most reform-oriented, has social norms that are operative for that particular class" (p. 148). These include the expectations that students are to explain their reasoning; agree or disagree with each other; and listen to, and make sense of other students' reasoning. When norms are evident in the classroom that are specifically related to the fact that the subject of study is mathematics, these norms become sociomathematical norms. Rasmussen, et al (2003) give an example of this distinction: "The expectation that one is to give an explanation falls within the influence of social norms, but what is taken as constituting an acceptable mathematical explanation is particular to the discipline of mathematics" (p. 150).

Although norms become routines through ongoing participation, they are interactively constituted and continually negotiated and renegotiated through interaction. When teachers begin the school year, the establishment of social norms is evident in classroom discourse and teacher-student interactions. But, as the school year progresses, these social norms become the building blocks for the sociomathematical norms that evolve in the mathematics classroom.

At the beginning of the school year, Cary guided her students through the development of community agreements. Although there were slight modifications each year depending on the particular groups of students, they typically included the following: be respectful listeners, always do your best, continue to try, ask questions, "three before me" (acknowledging that their classmates were also teachers), keep hands and feet to self, and wait to get drinks until independent work time. The "Golden Rule" was always "Treat others the way you want to be treated."

In addition, Cary modeled the types of questions students should ask each other when engaged in group work. She asked questions such as "I never thought of it that way," "Can you show me a different way using base ten pieces?" Her role as a teacher was to circulate amongst students, listen to their conversations, and reengage students in the mathematical task by asking a probing or extending question. During this time, Cary would monitor and select particular student work related to the core mathematical goal of the day's lesson and seek permission from students to share their solution strategies. She was selective regarding the students she chose at the beginning of the school year as they needed to be ones who could accept the challenges that arose in this situation. Because students were often not used to sharing their work with the class using a document camera, and felt disequilibrium towards these learning opportunities, Cary spent a great deal of time reframing this angst into celebrations of learning. At the beginning of the school year, Cary did not shy away from incorrect solutions or strategies. Instead, she used these as learning opportunities. When this occurred, the students worked together as a class to flesh out possible solutions using multiple strategies. They supported the presenting student by asking questions that showed an understanding of the thinking and led to a solution that had increased mathematical justification.

Moving students beyond just explaining their thinking or solution strategies to providing justification of their solutions using sound mathematical reasoning was an important step in developing sociomathematical norms in Cary's classroom. Cary began by asking students, "Can you prove that?" This was often followed up with the question, "How could you teach it to a second grader?" Regardless of whether the solution was correct or not, Cary would often act as the skeptic and ask them, "Are you sure you are right? How do you know you are right?" Prompting students to use manipulatives or sketches of their solution strategies would stress the importance of multiple representations. While selecting students' work for sharing, Cary would often look for multiple representations of the same strategy. Her goal then became to help students make connections and build upon the strategies for a more complete mathematical understanding of the concept.

During one of Higgins' first site-visits in Cary's classroom, she noticed that Cary had a poster on the wall with the following three questions: How did you solve the problem? Why did you solve it this way? Why do you think your solution is correct and makes sense? Classroom observation data indicated that students were going in and out of these questions with confidence, but often did not appear to be actively listening to each other. Higgins' field notes after that visit recorded her discussion with Cary. One of the things Cary was going to work on was "getting students to think about the questions in the room, but ask them to use more active listening—other questions from the moment, questions that were more authentic in nature."

After this visit, Cary had a class meeting regarding Higgins' comments and asked her students what they could do to change the nature of the questions so they were more authentic. This discussion proved to be extremely beneficial in raising their awareness of the types of questions they were asking. To reinforce what good listening "looks like," Cary changed the focus from explaining their own thinking to explaining the thinking of others in their group, and eventually to the student who was presenting his or her solution strategy during the plenary of the lesson. Again, Cary modeled by asking questions such as, "I don't understand what you just said," "Can you prove it to me a different way?" "I'm very visual. Can you use the manipulatives to show me?" "Yeah, but what if…?" Cary believed this last question, in particular, was pivotal in moving students into the realm of making generalizations and the establishment of sociomathematical norms in her classroom, as it forced students to move beyond a single case to a more general case.

The establishment of these sociomathematical norms continued during full-group discussions while Cary participated in the discussion at an empty student desk towards the back of the room pushing for the kind of mathematical reasoning she expected out of the learning community in her classroom. If a student was presenting, she would raise her hand, wait to be called on by the student, and probe and focus the discussion with a question once again that would raise the discourse and/or mathematical content to a higher level.

## CARY'S DISCOURSE-RICH CLASSROOM ENVIRONMENT

Stein (2001) claimed good tasks that elicit student thinking and discussion are an important first step in creating a discourse-rich environment. But, according to Stein, even with a good task, student discussions can fall into a rut and become predictable. Students' personal interest in a particular solution or strategy is a way that students can invest in and take ownership of the discourse (O'Conner, 1998, as cited in Stein, 2001). Stein believed that teachers have the "armament" to encourage this student interest by creating a classroom atmosphere of mutual respect and trust and selecting instructional tasks that prompt students to take different positions, find different solutions, and convince others of the correctness of solutions with mathematical evidence.

During OMLI, Cary spent time discussing the characteristics of cognitively rich tasks and ways to increase the cognitive demand of her tasks (Stein, Smith, Henningsen, & Silver, 2000). When she went back to her classroom, she looked at her curriculum differently in the sense that she thought about her lessons in light of the core mathematical ideas she needed to cover in her teaching. Rather than teaching each lesson as traditionally presented in the textbook, she became more selective and chose to structure the majority of her mathematics lessons around the most cognitively-demanding tasks. This was a challenge for Cary as the tasks had to be accessible to all levels of students, but at the same time press the students mathematically. As Cary got to know the mathematical abilities of her students she was able to choose seemingly simple problems that she knew would challenge the students and foster deeper mathematical understandings. The following episode is one example of this type of problem.

Cary's students were working on a fractions unit at the end of the school year and she wanted to extend their understandings of equivalent fractions. She asked students about the equivalency of two fractions: Does $2/3 = 10/15$? Does $2/3 = 20/30$? After each problem was presented and students had a few minutes of private think time, they engaged in discourse around the problem and justified their answers providing sound mathematical reasoning. To push students into disequilibrium and move them towards making a generalization, Cary asked the third question: Does $2/3 = 202/300$? After allowing time for small-group discourse, Cary focused the energy towards a large group discussion in which the students quickly took ownership. Here is an excerpt from the transcript of that lesson. A student, Lewis, is in the front of the room.

> Lewis: Me and Mark think it's correct because 2 is a factor of every even number.

(Students raise hands, with some rumbling it does work, and others that it doesn't.)

Lewis: [calls on student] Kate?
Kate: But we think if you times it...remember how we had times 5 in the last problem and then times 10 and it would work for each one. Do you think it would work here?
Mark: [points to board and pushes Lewis out to the way]
But I know that...I know that is right.... because 3 a hundred times ... so 100 3 times.
Kate: Yeah, but would it work with 2? In the other one ...we said it worked and we proved it...and it was times that same number and if it doesn't have that same number
how do we know it is correct?
Lewis: Because it has another number here [points at the 2 in the ones place of 202]. It's not just a zero.
Kate: Yeah, how do you know that it is not times... like something else, like 101? How do you know it is not times that?
Lewis: Because...[Mark is mentally multiplying and using his fingers to check]
Kate: So can you prove it?
Lewis: 101, 101 times 2 equals 202.
Kate: Can you prove that your answer is correct?
Lewis: AHHHHH! [calls on Georgia]
[Georgia comes up to the board with Mark and Lewis]
Georgia to Kate: Kate do you not understand what this was? [points to 202 and picks up pen]
Kate: I am just saying...um.... that I kind of need proof to know that they are... like timesed by the same number.
[Georgia writes 100 × 3 = 300 and (2 × 100) + 2 = 202 on board while Kate is talking]
Cary: OH...that is really interesting what you wrote up there, Georgia.

The discourse continued for a few more minutes until class was over. After reviewing the transcript, the researchers realized that three of the students engaged in this exchange were very reluctant mathematicians at the beginning of the year. Cary reflected on how the learning community in her classroom had become so inclusive that all students, regardless of their level of mathematical understanding, felt comfortable arguing with each other and discussing their thinking in front of the entire class. This contrasted greatly from how she needed to carefully select students to share at the beginning of the school year.

The researchers were interested in how students viewed their learning of mathematics in Cary's classroom compared to their previous two years in more traditional classrooms. Here are three students' responses:

> It's different because last year we were in rows so we couldn't talk in groups or we couldn't work together and hear their ideas because we would start talking about other stuff and not talk about math. So we had to sit in rows. But this year we sat in groups and it is so much easier.

> Math is different because we don't get as much help from the teacher telling us the answer. It is harder because we are learning geometry and fractions. On the other hand, this is the same because we still do adding and subtracting. In fact, math changes every school year. For example, my first grade teacher told me what 2 + 2 equaled, but my learning on my own makes it a lot easier because I get time to finally understand but if I get told, it is not fun and I forget because I am a visual.

> Learning math is different in third grade than first and second grade because in first and second grade we just had to do our work and turn it in. In third grade we would talk about the answers we got and how we know it's the right answer. The similarities in math in third, first, and second grade is that we learned how to multiply, add, subtract, and divide.

These three students give some indication of how they viewed elements of the mathematical environment Cary established in her classroom: working in groups, talking about mathematical ideas, explaining and justifying answers, mathematical sense making residing in the student rather than the teacher, and accessing multiple learning styles.

In Cary's discourse-rich mathematical environment, students are problem-solvers who often have multi-access points to any task presented. That belief system carried over to their attitudes and perseverance in mathematics, which Cary was able to observe as she monitored students in her school while taking the state mathematics benchmark test in the spring. Her students' mathematical power was evident in that they attempted any problem they had in front of them without assistance, in a manner that was calm and engaging. This was in contrast to other students in Cary's school from more traditional classrooms who experienced anxiety and often panic with the problems that were in front of them. Their lack of self-confidence in mathematics often paralyzed them, causing them to shut down. Cary's students had become the mathematicians she had envisioned!

## CONCLUSION AND DISCUSSION

According to Chapin and Eastman (1996), the learning environment and classroom culture that a teacher establishes has a tremendous influence on students' attitudes towards mathematics and the pursuit of knowledge as well as their intellectual and social development. "Constructing learning environments that develop communities of learners is more complex than simply encouraging discourse and asking students to complete mathematical tasks" (Chapin & Eastman, 1996, p. 115). They further believed that teachers themselves must be learners, thinkers, and risk takers. Cary possessed all these characteristics which contributed greatly to her success in creating a discourse-rich learning environment that led to powerful problem solvers, learners, and even teachers of mathematics.

The researchers hope the stories they have told of Cary and her classroom encapsulate the vision of reform mathematics as advocated by NCTM (1989, 1991, 2000), so that others will be able to learn from what she has accomplished, reflect on that learning, and take risks in their own classrooms.

The researchers next will go back to look at the original three research questions that guided this study and highlight their findings by pulling out key aspects of Cary and her teaching related to the questions. We acknowledge there is a risk in doing this by reducing rich stories into bullets of practice. But, we believe there is value in this process for others who would like to follow in Cary's footsteps.

## 1. How Does a Teacher Begin to Learn about Using Discourse Purposefully to Improve Mathematics Teaching and Learning?

The researchers found the following dispositions and experiences important to understanding discourse and implementing it purposefully in the mathematics classroom. Teachers need to:

- Desire to grow in their teaching of mathematics and be willing to take risks;
- Experience being the student in a discourse-rich environment;
- Collaborate with another colleague at the same grade level, using a lesson-planning framework to plan a lesson, co-teach the lesson in each other's classrooms, reflect on and fine-tune after the first teaching episode, and co-teach in second classroom;
- Observe another teacher to learn new questioning strategies and see students' cognitive responses to the questions being asked;
- Use protocols for collecting discourse data while observing another teacher and have a process for categorizing and analyzing the data. Analyze discourse through the lens of the core mathematical goal of the lesson;
- Whenever possible, involve a team of teachers, principal, and a mathematics coach in observations with a focus on student thinking and discourse.

*2. How Do Norms of Mathematical Discourse Evolve and How Can Teachers Influence the Development of These Norms?*

The researchers recommend teachers engage in the following processes and activities to help build the norms of mathematical discourse in their classrooms:

- Build social norms at the beginning of the school year through community agreements;
- View and honor disequilibrium as celebrations of learning and learning opportunities;
- Use protocols to establish routines for productive small-group discussions in order to minimize status and equity issues;
- Model questions for small-and-large group work that pushes students from just explaining their thinking to justifying their strategies and solutions using sound mathematical reasoning; press for generalizations whenever possible ("Yeah, but what if…?");
- Monitor, select, and sequence student work to be presented, based on core mathematical goals of lesson;
- Use incorrect solutions or strategies as opportunities for learning and the collaborative development of ideas; and
- Model and stress the use of authentic questions with students.

*3. What Does a Discourse-Rich Classroom Look Like and How Is It Different from More Traditional Classrooms?*

At the beginning of this article the researchers described Cary's classroom in a way that gave the reader a sense of what a discourse-rich classroom looked like. Although many classrooms could be considered discourse-rich in the sense that they use the strategies presented in this paper, something magical happened in Cary's classroom that was witnessed by the researchers. They would like to paint a picture of Cary's classroom the last week of school when she was able to step back and say to her students, "We're there! We've become a community of mathematicians, and it is beautiful!"

Cary presented the task, giving the students private think time, and then allowed them about 15 minutes to work through their disequilibrium in small groups. There was an energy and electricity in the air with all the minds working towards a common mathematical goal, not leaving anyone behind, and being respectful of everyone—their learning styles and processing speed. When Cary began the plenary after monitoring, selecting, and sequencing students' work to be presented to the class, as if nearly on cue, light bulbs began shooting off everywhere as the students and Cary probed and asked questions that deepened their understanding of the core mathematical goal of the task. Physically, the students eyes popped open, smiles formed on faces, and hands shot up. Students' actions demonstrated a burning desire to share their newly discovered strategies or to solve the problem by asking the next question that would take the class further or challenge each other's ideas.

What can we glean from this classroom that makes it different from traditional classrooms as experienced by the researchers?

- Student energy is evident when engaged in tasks of high-cognitive demand.
- Math talk is rich in high-level vocabulary around important mathematical ideas.
- There is a classroom atmosphere of mutual respect and trust which allows students to take risks and challenge each other.
- Private think time is an established norm and students value the opportunity to learn from each other through discourse in large-and-small group settings.
- Students take ownership of their own and each other's learning as well as their understanding and sense making of mathematics.
- The classroom is an inclusive learning community where all students' perspectives are honored and the teacher is viewed as a co-contributor.

Although this paper is a case study of one teacher and her classroom, the researchers believe there are implications for the teaching and learning of mathematics that are important in all teacher training programs—inservice or preservice It is imperative that professional development leaders and teacher educators model and explore ways to bring the experiences of mathematical discourse into their own teaching so that the spirit of mathematics reform can come alive within the minds and hearts of teachers. Only through this experience will teachers have the capacity and desire to bring it back to their own classrooms and unleash the mathematician in each one of their students!

## REFERENCES

Ball, D. L., & Bass, H. (2003). Making mathematics reasonable in school. In Martin, G. (Ed.), *Research compendium for the Principles and Standards for School Mathematics* (pp. 27–44). Reston, VA: National Council of Teachers of Mathematics.

Ball, D.L. (1991, November). What's all this talk about 'discourse'? (Implementing the Professional Standards for Teaching Mathematics department). *Arithmetic Teacher, 39* (3), 44–48.

Clandinin, D. J., & Connelly, F. M. (2000). *Narrative inquiry: Experience and story in qualitative research*. San Francisco, CA: Jossey-Bass.

Chapin, S. H., & Eastman, K. E. (1996). External and internal characteristics of learning environments (Implementing the Professional Standards for Teaching Mathematics department). *Mathematics Teacher, 89*(2), 112–115.

Foucault, M. (1972). *The Archaeology of Knowledge*. New York: Pantheon.

Franke, M. L., Kazemi, E., & Battey, D. (2007). Mathematics teaching and classroom practice. In Lester, F. K. (Ed.), *Second handbook of research on mathematics and teaching*. (Vol. 1, pp. 225–256). Reston, VA: National Council of Teachers of Mathematics.

Hiebert, J., Morris, A. K., & Glass, B. (2003). Learning to learn to teach: An "experiment" model for teaching and teacher preparation in mathematics. *Journal for Mathematics Teacher Education, 6*, 201–222.

Kazemi, E. (1998, March). Discourse that promotes conceptual understanding (Research into Practice department). *Teaching Children Mathematics, 4*(7), 10–14.

National Council of Teachers of Mathematics. (1989). *Curriculum and evaluation standards for school mathematics*. Reston, VA: Author.

National Council of Teachers of Mathematics. (1991). *Professional standards for teaching mathematics*. Reston, VA: Author.

National Council of Teachers of Mathematics. (2000). *Principles and standards for school mathematics*. Reston, VA: Author.

Rasmussen, C., Yackel,E., & King, K. ( 2003). Social and sociomathematical norms in the mathematical classroom. In Schoen, H. L. & Charles, R.I. (Eds.), *Teaching mathematics through problem solving, grades 6–12* (pp.143–154). Reston, VA: National Council of Teachers of Mathematics

Schwandt, T. A. (2001). *Dictionary of qualitative inquiry* (2nd ed.). Thousand Oaks, CA: Sage Publications.

Sherin, M. G. (2002). A balancing act: Developing a discourse community in a mathematics classroom. *Journal of Mathematics Teacher Education, 5*, 205–233.

Smith III, J. P. (1996). Efficacy and teaching mathematics by telling: A challenge for reform. *Journal for Research in Mathematics Education, 27*(4), 387–402.

Smith, M. S. (2002). Redefining success in mathematics teaching and learning. In Teppo, A.R. (Ed.), *Reflecting on NCTM's principles and standards in elementary and middle school mathematics: Readings from NCTM's school-based journals*. Reston, VA: National Council of Teachers of Mathematics.

Springer, G. T., & Dick, T. (2006). Making the right (discourse) moves: Facilitating discussions in the mathematics classroom. *Mathematics Teacher, 100*(2), 105–109.

Stein, M. K. (2001, October). Mathematical argumentation: Putting the umph into classroom discussions (Take Time for Action department). *Mathematics Teaching in the Middle School, 7*(2), 110–112.

Stein, M. K., Smith, M. S., Henningsen, M. A., & Silver, E. A. (2000). *Implementing standardes-based mathematics instruction: A casebook for professional development*. New York: Teachers College Press.

VanZoest, L. R., & Enyart, A. (1998). Discourse, of course: Encouraging genuine mathematical conversations. *Mathematics Teaching in the Middle School, 4*(3), 150–157.

# CHAPTER 5

# THE ROLE OF TASKS IN PROMOTING DISCOURSE SUPPORTING MATHEMATICAL LEARNING

## Sean Larsen and Joanna Bartlo

In this paper we consider the role that tasks can play in promoting substantive mathematical discourse that leads to opportunities for learning. We will describe a task that we have used successfully in an algebra course for K–12 in-service teachers as part of the Oregon Mathematics Leadership Institutes (OMLI)[1]. An analysis of both the task itself and the participants' mathematical activity as they engaged with the task will be provided with the goal of identifying and articulating: the kinds of mathematical discourse that were generated, the kinds of opportunities for mathematical learning that this discourse provided, and design features of the task that seemed to support the discourse and subsequent learning opportunities.

KEYWORDS: promoting discourse, learning, instructional design, tasks, symmetry, justification

[1]The Oregon Mathematics Leadership Institute Partnership Project is funded by the National Science Foundation's Math Science Partnership Program (NSF-MSP award #0412553) and through the Oregon Department of Education's MSP program..

# INTRODUCTION

Recent reform documents, such as those published by the National Council of Teachers of Mathematics (NCTM, 1991, 2000) have called for a shift from a focus primarily on procedural knowledge of mathematics to one that includes conceptual understanding. Making such a shift requires rethinking the learning goals we set for our students and how students can best achieve these goals (Hiebert, 2003). This means classrooms are shifting from systems where teachers are the purveyors of knowledge, to systems in which students take an active role in their own learning. To that end, communication is becoming an increasingly important component of classroom practice and researchers have become increasingly interested in studying the discourse that occurs in classrooms (Ilaria, 2002).

Students can learn about what it means to do mathematics by engaging in activities such as questioning, challenging, and justifying (Stein, 2007). They can also construct meanings for mathematical ideas (NCTM, 2000). For instance, by reflecting and building on mathematical explanations of their peers, students can reorganize their ideas and make connections between ideas (Forman, 2003; Hatano & Inagaki, 1991; Lampert, 1986). Additionally, as students communicate their reasoning, they can develop more sophisticated understandings of mathematical ideas (Simon & Blume, 1996). For example, as they communicate and justify conjectures, students engage in generalization and abstraction, processes that lead to more sophisticated mathematical understandings (Cobb, Boufi, McClain, & Whitenack, 1997; Lampert & Cobb, 2003).

Generating productive discussions requires more than simply asking students to talk. The mathematical discourse is shaped by the instructional tasks and by how the discussions are facilitated (Himmelberger & Schwartz, 2007). Since the quality of discourse affects the potential for mathematical understanding (Kazemi & Stipek, 2001), it is important to use tasks that can move students toward precision, clarification, and generalization (Himmelberger & Schwartz, 2007). Further, the discourse needs to be centered on meaningful mathematical ideas and be based on the ways students come to know those ideas (Kysh, Thompson, & Vicinus, 2007). It is important to ensure that all students participate in the discussions (e.g. Stein, 2007), and that students' contributions to discussions go beyond their personal ideas and connect to the ideas of others (Staples & Colonis, 2007).

In sum, discourse is seen to have the potential to play an important role in the learning of mathematics. However, it is important that the conversations focus on significant mathematical ideas and that the nature of the students' discourse supports the learning of these ideas. Furthermore, to support learning for *all* students, it is important that *all* students are active participants in the classroom discourse.

In this paper we consider the role that tasks can play in promoting substantive mathematical discourse that leads to opportunities for learning. We will describe a task that we have used successfully in an algebra course for K–12 in-service teachers as part of the Oregon Mathematics Leadership Institutes (OMLI). An analysis of both the task itself and the participants' mathematical activity as they engaged with the task will be provided with the goal of identifying and articulating:

1. the kinds of mathematical discourse that was generated;
2. the kinds of opportunities for mathematical learning that this discourse provided; and
3. design features of the task that seemed to support the discourse and subsequent learning opportunities.

The goal of this analysis is to generate design principles for developing tasks that promote rich mathematical discourse and to explicate the interaction between tasks and discourse in promoting learning.

## SETTING/METHODS

### The Setting

The OMLI algebra course consisted of 15 two-hour sessions over the course of three consecutive weeks. The section of the course from which we drew our data consisted of 16 students (usually seated in groups of four) and was co-facilitated by the first author and an experienced K–12 mathematics teacher and administrator with extensive experience facilitating professional development. Modes of instruction included work on tasks individually, in pairs, and in small groups as well as discussion in pairs, small groups, and whole class.

### Methods

Each session of the course was videotaped using two cameras to capture work in two of the four small groups as well as the activity at the front of the room and in the audience during whole class discussions. All of the students' written work was digitally copied and all of the posters created during class were digitally photographed. Additional data sources included instructional prompts, facilitation protocols, and notes regarding instructional design decisions.

The data corpus was analyzed using an iterative method derived from techniques described by Cobb and Whitenack (1996) and Lesh and Lehrer (2000). The first phase of our analysis consisted of viewing videotapes of the classroom activity to code the students' discourse using the OMLI Classroom Observation Protocol (Weaver, Dick, & Rigelman, 2005). The second phase of our analysis involved looking more closely at the justifications identified during the first phase of analysis. We analyzed the level of sophistication of the students' justifications and the nature of the bases (or foundations) of these justifications. We also attended to how these evolved throughout the task sequence. The third phase of our analysis focused on the mathematical content of the students' discourse. Specifically, we attended to opportunities for mathematical learning that resulted from the mathematical discourse, paying special attention to notions of equivalence as that was an important learning goal for the sequence. The fourth phase of our analysis focused on the task sequence itself in light of the earlier phases of analysis. Specifically, we worked to identify characteristics of the task sequence that seemed to promote the phenomena we had identified in our earlier analyses of the students' discourse.

## The Measuring Symmetry Task Sequence

In this section we provide the context for the analyses that follow by describing the first task of the Measuring Symmetry sequence, briefly describing the students' mathematical activity in response to this task, and describing a follow-up task that is the focus of much of the discourse we will consider. The Measuring Symmetry task sequence was the second instructional sequence of the course and was initiated on the third day of instruction. The starting point of the sequence was a task (Figure 1) that required the students to first rank six figures from least symmetric to most symmetric and then create a system for measuring (assigning a number to) the symmetry of any figure. The students were asked to do this by relying on their intuition and aesthetic sense.

This task seemed to generate a tremendous amount of substantive mathematical discourse among the students. After having approximately ten minutes to work individually on the first prompt (ranking the figures), the students took turns explaining their thinking to the other members of their small groups. These initial explanations revealed similarities and differences in approaches that were reflective of the different access points afforded by the task (e.g. some focused exclusively on bilateral symmetry while others included rotational symmetry in their considerations). Some students anticipated the second prompt and ranked the figures according to the number of symmetries (reflection and/or rotational) they had. Others used more informal ranking systems. Additionally, individuals varied according to whether they made an effort to break ties and in how they did

1. Order the figures below from least symmetry to most symmetry.

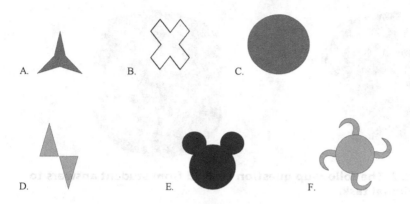

2. Create and describe in detail a method for measuring a figure's symmetry.

**FIGURE 1. The Measuring Symmetry task.**

so. When the second prompt was revealed and the students worked to come to consensus on a system for measuring a figure's symmetry, they began to engage in more substantive discourse including generalizing and justifying. Much of the justification activity was focused on determining how to handle rotation symmetries. Various students offered up and justified approaches with arguments based on their intuition and aesthetic sense (which is consistent with the instructions given in the task). In this way, this initial task provided a safe context for mathematical argumentation (since these arguments were based on intuition other students could disagree with them without undermining their confidence).

The students' discourse around this initial task touched on a number of significant mathematical ideas with which we hoped to engage the students. For instance, the initial discourse raised the question of whether the notion of symmetry goes beyond the idea of bilateral symmetry. It also set the foundation for thinking of symmetries as objects (because the students were essentially counting them). Finally, and most relevant for the discussion that follows, the students' work set the foundation for exploring the notion of equivalence in general (and equivalence of symmetries in particular).

Much of the discourse we will consider in the next two sections was focused on a follow-up task that we designed for the second day of the measuring symmetry sequence. During the first day of the sequence, each group of students developed a poster illustrating their ranking of the fig-

Consider the following poster scraps. How would you respond to the following question: "Each of these say that the figure has rotational symmetries, but aren't there really 6 different rotational symmetries depicted?"

**FIGURE 2. The follow-up question created from student answers to the original task.**

ures and their method of quantifying a figure's symmetry. We took digital photographs of the posters and the next day posed a question (Figure 2) that compared aspects of two different systems (one that counted counterclockwise rotations and one that counted clockwise rotations). We asked whether the depicted figure had three rotational symmetries as shown on each poster, or if it in fact had six rotational symmetries (three in each direction).

We will provide a detailed description and analysis of the students' mathematical discourse around the Measuring Symmetry task and this follow-up task. This will feature an analysis of a whole class discussion that featured diverse and increasingly sophisticated mathematical discourse. We follow this with an analysis of the mathematical content of the students' discourse, explicating the opportunities for learning about equivalence that resulted from the discourse. Finally, we examine the Measuring Symmetry sequence itself in an effort to identify characteristics that seemed to promote the discourse that we observed and the resulting opportunities for learning.

## RESULTS OF OUR ANALYSES

### The Students Mathematical Discourse

The first phase of our analysis consisted of viewing videotapes of the classroom activity to code the students' discourse using the OMLI Classroom Observation Protocol (Weaver et al., 2005). This analysis revealed that the students engaged in a wide variety of types of discourse. Further analysis revealed that the sophistication of the justifications increased throughout the engagement with the task. In this section we will elaborate the varied types and levels of discourse the students engaged in while working on the task.

### The Original Task

Since the first prompt asks the students to make a decision based on their intuition, the mathematical discourse began with students sharing their rankings. The students often explained their rationale for their rankings while describing them. Since the students used different systems for ranking the images, the early discourse also involved questioning from group members who were trying to understand their peers' reasoning. This was usually followed by a response, which often included an explanation or justification. The following group discussion excerpt illustrates this type of interaction.

Ruth was the first person in her group to share her ranking, which she constructed considering only reflection symmetry. Daniel shared next, and he included both rotational and reflection symmetry. The following excerpt picks up with the discussions of Daniel's ranking.

> Daniel: Also if you fold it in half—if you fold it this way, and you fold it this way, and you can rotate this one…I used rotational symmetry too. So three lines of symmetry plus three points of rotational symmetry.
> Ruth: Explain that one to me.
> Daniel: If you take this point, and rotate around it, it would be the same picture.
> Ruth: By changing what you are doing.
> Amber: I put my pencil down, and I turn it, I get here, and I get here again, so it's the exact same figure. You probably can't turn Mickey so he lands back on.
> Daniel: You can turn him all the way around, so it has at least one.

This discussion began with Daniel sharing his ranking, which included an explanation of how he chose his ranking. His explanation included ideas with which Ruth was unfamiliar, so she asked questions to understand Daniel's reasoning. Both Amber and Daniel offered answers to Ruth's questions by explaining their understanding of rotational symmetry. Daniel then challenged Amber's claim that "Mickey" does not have any rotational symmetry, and justified his challenge by arguing that this figure has a 360-degree (or full turn) rotational symmetry. By basing the discussion on their personal ideas about symmetry, the students were able to comfortably share their ideas as well as explain and justify them.

As the group's engagement with the task progressed, and they worked to come to a consensus, the number of justifications and challenges increased.

For instance, while the students tried to decide on the order of the images, they began to challenge each other's rankings. Sometimes the challenges included a justification for the challenge, and sometimes justifications were offered as students rebutted the challenges. In the example that follows both types of justifications occur.

Amber: We are going with Mickey.

Ruth: Because of the line of symmetry.

Amber: And we are giving the line of symmetry a higher priority.

Daniel: But he only has one line of symmetry and this one has three rotations.

Amber: I know, but we are still going to give it a higher priority, because people see it better. So we have 3 lines, 2 lines, 1 line, and then no lines.

At this point the students were beginning to engage more in argumentation than in sharing their opinions. However, their justifications were still largely based on their opinions about symmetry.

In response to the second prompt the students began to generalize beyond specific examples. In this way they added another type of discourse to their discussions, which caused the discussions to increase in mathematical sophistication. The following excerpt illustrates this shift to thinking in general terms rather than about the particular images from the task.

Abby: What about if we give an aesthetic scale, it is it a 1, 2, or 3?

Amber: What if we had one figure in one room and another figure in another room, am I able to compare them, am I able to assign an aesthetic value to one of those figures? I don't know. I don't think so. I'm able to measure the symmetries and give it a numerical value, but aesthetic is hard.

Abby: I understand what you are saying, but that is not how we set up this table, we said Mickey Mouse is fun. That's not quantified yet.

Amber: But we added a quantifiable reason for it, because it has lines and not rotations.

In this discussion the students moved from talking about how symmetric these shapes were, to what made a figure more symmetric in general (which was necessary in order to develop a method for quantifying the symmetry of a figure). Note that their initial consideration of generalization comes in the form of challenges and justifications. Therefore, through the process of challenging and justifying the students began to generalize.

### The Follow-Up Task

A similar diversity of types of discourse arose during the follow-up task. The discussions often took a similar form to those around the first task, but with more of an emphasis on explaining, challenging, and justifying. One example of this pattern can be seen in the following whole-class discussion excerpt.

> Penny: You're rotating clockwise in the first figure. And you're rotating counterclockwise in the second figure. So you have a positive angle of rotation in the first one and a negative angle of rotation in the second one.
>
> Susan: So what would you call that?
>
> Penny: I would call it six.
>
> Susan: You would call it six?
>
> Penny: I would call it six.
>
> Kathy: Well I count them differently. I don't count them as the number of turns, I think of them as degrees also. But for example on the first one, if you do that twice—
>
> Instructor: Which one?
>
> Kathy: The one on the left, if you turn that one twice and I numbered them I put point 1, point 2, point 3. Then where point one is located and you take the one on the right and you only turn it once to the left, those two pictures are exactly the same. So moving this one twice, the one on the left, twice and this one back once, the orientation of the picture is exactly the same. So those two rotations basically are the same thing. Since the position of the points are in the same location and you've double counted. You're counting positions twice.
>
> Susan: So it's kind of like whether you're counting the motion as the thing or the ending position as the thing.
>
> Kathy: Right and I'm saying they overlap each other.

In this excerpt both Penny and Kathy share their opinions about the question, and they accompany their answers with justifications. Kathy's answer, and accompanying justification, was presented as a challenge to Penny's response. Susan asked questions to help clarify her understanding of both arguments.

The discussion of whether the motions are important (or only the results) morphed into a discussion of different types of equivalence. This conversation also featured a diversity of types of discourse, but justification began to be built on earlier class discussions and other mathematical ideas rather than on opinions.

Sophie: 24, The number 24, you can say 2 × 12 it would get 24 and you can say 4 × 6 and 2(3) × 8 and they all come up with the same answer. And we're all comfortable with the fact that there's several different ways, several different paths, to get the same answer. So ending result is the same but we count all these problems kind of differently. And the sense that like we were saying negative and positive turns right, in the end result is the same but you took a different path and is that ok?

Jim: But if we tie back into our original definition that we've agreed on, then it's just a movement that changes a shape. Then if you're saying the movements OK then well I'm gonna move, I mean if you look at this (holds up two square post-it notes) ok we'll limit it to this this this or this this this (Shows turns in each direction), well here's a movement that changed it and now it's back to its (moves the figure in a large motion lifting one square a foot or so above the other and wiggling it around before bringing it back down to rest on the other square) well watch this one (Does even more complicated motion) (Laughter) and it gets into the infinite, I don't want infinite symmetry.

Instructor: That's a great point. That's why we have to decide, that some of these things are the same as other ones. So we need to pin down what do we mean by equivalent.

Amber: Well I think Sophie's example doesn't fit my idea of equivalence as much as saying 2 × 4 and 4 × 2.

Sophie begins this discussion by sharing her opinion and relating it to something familiar to everyone in the classroom. Jim then challenges Sophie's idea, and justifies his challenge by relating his response to earlier class discussions and by modeling an illustrative example of his idea for his peers. Amber also challenges Sophie's statement, and justifies her opinion by using a mathematical example that is related to Sophie's.

### Bases of Justifications

Although the students engaged in justifications throughout the entirety of the task, the justifications changed form throughout the task because the students began to base their justifications on previous discussions. In this section, we explicate the different bases the students used as they progressed through the task.

In the initial part of the task students based their decisions on their opinions and intuition. This helped students feel confident in their answers, and enabled them to justify their answers. This meant, however, that the justifications were based entirely on the students' beliefs or opinions. An example of this is when Amber said "I know, but we are still going to give it a higher priority, because people see it better." In this case she was basing her argument on her beliefs about people's views of line symmetry versus rotational symmetry.

As the small groups came to a consensus on their rankings, they were then able to use those agreements as bases for their justifications. When Abby said, "I understand what you are saying, but that is not how we set up this table, we said Mickey mouse is fun. That's not quantified yet," she was basing her justification on the consensus her group had reached earlier. In this way her justification was no longer based on opinions or beliefs, but on something that had been established by the group.

Once again, the agreements reached from the argumentation described above moved on to become the bases of future justifications. Eventually the students created working definitions upon which they could base their justifications. This can be seen when Jim said "but if we tie back into our original definition that we've agreed on, then it's just a movement that changes a shape." In this way, Jim used the class's emerging definition of symmetry to justify his view of equivalent symmetries.

The students' justifications continued to be based on increasingly sophisticated ideas. For instance, as the class worked on defining equivalent symmetrics, Kathy suggested that two motions could be considered equivalent if all of the points of a figure end in the same location after each motion is applied to the figure. The students were then able to use Kathy's definition to justify their answers to later questions. Jim did this when he argued that no reflection could be equivalent to a rotation by saying "No, because one of the points is still in the same spot, but the other two have shifted." In this way the students began to engage in more sophisticated and rigorous justification as they began to base their justifications on emerging mathematical definitions.

Above we saw that the variety of types of discourse the students produced expanded throughout their engagement with the task. Here we saw that not only the variety of types of discourse increased, but the sophistication increased as well. That is, the students went from basing their arguments on their opinions to basing their arguments on more mathematical foundations.

### Opportunities to Learn about Equivalence that Resulted from the Students' Discourse

Equivalence is one of the most important ideas at all levels and in all areas of mathematics. Equivalence was a central theme of the OMLI Algebraic Structures course and we explored notions of equivalence that spanned the K-16 grade range. At various times we considered equivalent arithmetic expressions (e.g. $2 + 2$ and $1 + 1 + 1 + 1$), equivalent algebraic expressions (e.g. $a(b + c)$ and $ab + ac$), and equivalent algebraic structures (e.g. group isomorphism). The notion of equivalent symmetries is somewhat paradigmatic in the sense that it involves focusing on the sameness of the result of a process rather than the process itself. This idea underlies many examples of equivalence (including equivalent arithmetic and algebraic expressions) and is captured formally in the definition of equivalence of functions. Two functions are equivalent if they always produce the same output given the same input, and of course equivalence of symmetry transformations is a special case of equivalence of functions. We designed the Measuring Symmetry sequence in part to support our overarching goal of enriching the students' conceptions of equivalence. In this section, we take another look at the whole class discussions analyzed in the previous section, this time attending to the opportunities for learning about equivalence that were afforded by the students' discourse.

### Tension Between Attending to Sameness and Attending to differences

The first phase of the whole class discussion featured a back and forth between students arguing for the differences between various symmetries and students arguing for the sameness of various symmetries. Penny led off with an argument that clockwise rotations are different than counterclockwise rotations. Penny's argument was based on the fact that the motions felt different to her.

> Penny: You're rotating clockwise in the first figure. And you're rotating counterclockwise in the second figure. So you have a positive angle of rotation in the first one and a negative angle of rotation in the second one.
> Susan: So what would you call it?
> Penny: I would call it six

Kathy then responded by arguing that a 240-degree clockwise rotation and a 120-degree counterclockwise rotation are the same because they change the orientation of the figure in the same way. Kathy's argument was based on the fact that she felt that it was the ending position that matters and not the motion.

> Kathy: Well I count them differently. I don't count them as the number of turns, I think of them as degrees also. But for example on the first one, if you do that twice —
> Instructor: Which one?
> Kathy: The one on the left, if you turn that one twice and I numbered them I put point 1, point 2, point 3. Then where point one is lo-

cated and you take the one on the right and you only turn it once
to the left, those two pictures are the exactly the same. So moving
this one twice, the one on the left, twice and this one back once,
the orientation of the picture is exactly the same. So those two rotations basically are the same thing. Since the position of the points
are in the same location you've double counted. You're counting
positions twice.

Susan: So it's kind of like whether you're counting the motion as the
thing or the ending position as the thing.

Note that Susan made an important contribution by pointing out that
Kathy's approach (which argues for a weaker condition for equivalence)
involved attending to the effect the motion has on the figure rather than
merely the motion itself, whereas Penny's approach (which argues for a
very strict condition for equivalence) involves attending almost exclusively
to the actual motion. As a result the debate shifted slightly, but significantly,
to one in which arguments were given either in support of or against the
idea of focusing on the effect of the motion. This allowed the students to
shift from focusing on their beliefs to relating their arguments to other
mathematical ideas.

The discussion began with Sophie arguing that in arithmetic different
multiplication problems are considered to be different even if they produce the same answers.

Sophie: 24, The number 24, you can say $2 \times 12$ it would get 24 and you
can say $4 \times 6$ and $2(3) \times 8$ and they all come up with the same answer. And we're all comfortable with the fact that there's several
different ways, several different paths, to get the same answer. So
ending result is the same but we count all these problems kind of
differently. And the sense that like we were saying negative and
positive turns right, if the end result is the same but you took a different path and is that ok?

Jim then brought the discussion back to the original task (ranking/mea-
suring symmetries) and to the class's definition of symmetry, arguing that if
one focuses only on the motion when determining equivalence then all of
the figures have infinitely many symmetries (which would of course render
the measurement system meaningless).

Jim: But if we tie back into our original definition that we've agreed on,
then it's just a movement that changes a shape. Then if you're saying the movements ok then well I'm gonna move, I mean if you
look at this (holds up two square post-it notes) ok we'll limit it to

this this this or this this this (shows turns in each direction), well here's a movement that changed it and now it's back to its (moves the figure in a large motion lifting one square a foot or so above the other and wiggling it around before bringing it back down to rest on the other square) well watch this one (does even more complicated motion) (laughter) and it gets into the infinite, I don't want infinite symmetry.

### Articulating and Arguing with an Equivalence Criterion

Following the whole-class discussion, the students were asked to develop a system for discerning whether two symmetries were equivalent. Kathy presented her approach to the class.

Kathy: My triangles have the different numbers. So here is my starting one with the 1, 3, and 2. And if I rotate it a 120 my 3, 2, and 1 move and if I rotate 240, they move again. So a 120 and a 240 and a 360 are all three different orientations. But 360 and 720 they're the same orientation. So it takes care of Jim's problem where 360 and 720 and everything, they're all the same. If you go with going the other direction and you go a negative 120. Notice a negative 120 and a 240 have the same orientation so they're considered the same symmetry. But the 240 and the 120 are different because the numbers are in a different location. Does that help?

Then one of the instructors asked the class whether they could use Kathy's idea to determine whether a reflection (a flip across the vertical in particular) could be equivalent to a rotation. Immediately Jim and Susan (among others) said this reflection could not be equivalent to a rotation. Jim quickly gave a justification using Kathy's working definition of equivalence, arguing that a reflection cannot be equivalent to a rotation because the ending orientation of the figure will necessarily change in a different way than it would under a rotation.

Jim: No, because one of the points is still in the same spot, but the other two have shifted.

In this discussion Jim (and other students) based their argument on the emerging definition of equivalence (based on Kathy's idea) rather than on their own opinions about equivalence. In this process the class began to make sense of Kathy's idea, which helped them begin to evaluate it. This was an important part of the process of creating a class definition of equivalence and the students' learning of this concept.

*Summary*

As they engaged with the measuring symmetry task (and particularly the follow-up question), the students engaged in substantive discourse around the issue of equivalence. By beginning with their intuitive notions about equivalence and extending to the class' working definitions, this discourse led to a tension between attending to similarities among symmetries (in particular the impact they have on the figure) and attending to differences between symmetries (in particular the differences between the actual motions performed). This tension created an intellectual need to develop a notion of equivalence that could resolve the tension in a way that was consistent with the students' informal notions of symmetry as seen in their initial measurement schemes. To do this the students needed to stop relying exclusively on their intuition about the task, and to base their arguments more on class consensuses and working definitions, which had evolved from their intuition and mathematical explorations. Thus the task both created a need to consider the issue of equivalence and a context for supporting mathematically meaningful arguments for and against proposed notions of equivalence (an hence support for formulating a definition of equivalence.)

## Characteristics of the Measuring Symmetry Task Sequence

In this section, we will analyze some aspects of the measuring symmetry sequence with an eye toward teasing out those characteristics that supported the productive mathematical discourse we observed and the learning opportunities that resulted from that discourse. The characteristics described here emerged as a result of coordinating our analysis of the students' discourse with our analysis of the resulting opportunities for learning. These analyses were significantly influenced by our awareness and experience with several frameworks related to instructional design. In particular, we will draw connections between our observations and the models and modeling perspective (Lesh, Cramer, Doerr, Post, & Zawojewski, 2003), Harel's (2007) DNR framework, and Rasmussen and Marrongelle's (2006) notion of a pedagogical content tool.[2] Relevant aspects of these frameworks will be elaborated in the context of explicating our observations.

In the following subsections, we elaborate four characteristics of the measuring symmetry task that seem to have contributed to what we observed in the students' mathematical discourse:

1.    The task sequence provided multiple access points for the students.

---

[2]The more general instructional design theory of Realistic Mathematics Education (Gravemeijer, 1999) also influences our analysis and is related in many ways to the more specific constructs we describe.

2. The task called for the creation of a mathematically significant model.
3. The task (and the follow-up question) admitted multiple *justifiable* approaches.
4. The task provided a context that could be leveraged to bring the argumentation to a resolution.

## The Tasks Provided Multiple Access Points

The initial measuring symmetry task was one that was accessible to all of the students. All of the students had some aesthetic sense of symmetry that they relied on as they formulated their responses, while some of the students could approach the task more formally based on their experiences. For example some of the elementary teachers mentioned having taught bilateral symmetry to their students. However, even without those kinds of experiences, students could make progress on the task by analyzing the figures in an effort to quantify their intuitive sense of symmetry. Regardless of how the students approached the problem they had an answer to share and explanations for their answers.

This characteristic of the task is in line with the *reality principle* from the models and modeling perspective (Lesh et al., 2003). The reality principle is one of the principles for designing what Lesh et al. refer to as model-eliciting activities. The reality principle asks the question: Will students make sense of the situation by extending their own knowledge and experiences? In the Measuring Symmetry sequence the students relied on their own aesthetic sense and experiences (formal or otherwise) with symmetry in order to produce their rankings and measurement systems. The task had the characteristic that it could be approached using whatever prior knowledge or experiences the individual students brought to the task. And the variety of responses that resulted provided a rich context for the students' discourse. That is, it provided each student an entry point to the discussion.

### The Tasks Called for the Creation of a Mathematically-Significant Model

Another of the principles for the design of model eliciting activities described by Lesh et al. (2003) is the *model construction principle*. An activity satisfies this principle if it immerses the students in a situation in which they are likely to confront the need to develop (or refine, or modify, or extend) a mathematically-significant construct. In the case of the measuring symmetry sequence, the students are asked to develop a method for quantifying the symmetry of a figure. In order to create such a system, the students needed to consider a number of questions either explicitly or implicitly:

- Do rotational symmetries count?
- If rotational symmetries count, should they be weighted less than reflection symmetries?
- Does a 360 degree or 0 degree rotation count?
- When should two symmetries be considered to be the same?

In order to answer these questions and produce their final quantification scheme, the students needed to debate with each other. This means that students needed to ask questions to understand each other's ideas, explain their ideas, challenge their ideas, and justify their ideas. In this way, this characteristic of the task seems to be largely responsible for the patterns of discourse and opportunities for learning that we observed.

Additionally, the model construction principle requires that the construct the students are asked to produce be mathematically significant. In the case of the measuring symmetry task, the model the students are asked to produce must account for a number of significant issues. In particular, a method for quantifying the symmetry will depend on a method for determining whether two symmetries are equivalent. In this sense the task creates what Harel (2007) refers to as an *intellectual need* to determine what it means for two figures to be equivalent. The follow-up task served to make this need more pressing by confronting the students with a discrepancy in how two groups were counting symmetries that could only be resolved by developing a notion of equivalence. So the measuring symmetry sequence calls for the creation of the very mathematically significant construct of equivalence of symmetries. This pressed the students to base their justifications on increasingly significant ideas. The combination of these characteristics seemed to contribute significantly to the opportunities for learning that we observed.

### The Tasks Admitted Multiple Justifiable Approaches

The initial tasks in the Measuring Symmetry sequence could be approached in any number of different and justifiable ways. For instance when ranking figures, it is reasonable to include a 360-degree rotation as a symmetry (which some groups did). However, it is also reasonable to exclude a 360-degree rotation. Furthermore, these different approaches are mathematically justifiable. For instance, students argued that a 360-degree rotation is much like a zero in addition (in that it does not change the figure) and so this rotation should be included. On the other hand, students argued that every figure has a 360-degree rotation and so including this rotation artificially inflates the symmetry measurement of every figure and as a result the measurement method does not accurately capture the idea of having no symmetry.

The follow-up task also admitted more than one justifiable response. As we saw above, students were able to provide convincing arguments that one should not consider a 240-degree clockwise rotation to be the same as a 120-degree counterclockwise rotation, while others were able to argue convincingly that one should indeed consider these to be the same. Clearly this characteristic of the task (and the follow-up) helped to set the stage for the mathematical argumentation that we observed by allowing for two opposing positions to be considered and justified.

In this way, this follow-up task seems related to what Rasmussen and Marrongelle (2006) call a *generative alternative*. "Generative alternatives are defined as alternate symbolic expressions or graphical representations that a teacher uses to foster particular social norms for explanation and that generate student justifications for the validity of these alternatives." (p. 389). In this case, we used the two ways of thinking about rotations (clockwise and counterclockwise) to pose two alternative ways to count these (as three rotations or as six rotations). And as we have shown above, the students did indeed provide justifications for each of these alternatives. And these justifications were crucial to the eventual development of the notion of equivalence of symmetries.

### The Tasks Provided a Context that Supported the Resolution of the Argumentation

While we certainly see the intrinsic value of having students engage in mathematical discourse, we also wish to emphasize that an important desired outcome of mathematical discourse is the learning of specific mathematical content. The first three characteristics of the measuring symmetry sequence that we have elaborated all seem to have contributed to the generation of significant mathematical discourse on the part of the students, setting the stage for learning about equivalence. However the task had another feature that seemed to be particularly important for moving the students' thinking forward. The task provided a context for resolving the mathematical arguments that it generated. In this case, the students needed to make a decision to consider different symmetries to be equivalent if they were to develop a system for measuring symmetry that was mathematically meaningful. Additionally, this decision needed to be based on earlier arguments and decisions. We saw this in Jim's argument that closed the whole class discussion by appealing to the emerging definition of symmetry and the goal of meaningfully measuring symmetry. This characteristic is consistent with the *self-evaluation* principle described by Lesh et al. (2003), in which an activity should promote self-evaluation on the part of the students. The measuring symmetry context provided criteria for the students to evaluate their stance on whether to focus on the similarities or differences between symmetries. This feature cemented the intellectual need to develop a notion of equivalence that focused on the result of the symmetry and not on the actual motion performed.

## CONCLUSIONS

In this paper we presented an analysis of a group of students' (teachers in this case) discourse as they engaged with the Measuring Symmetry sequence and examined the opportunities for learning that participating in this discourse afforded them. We also examined the task sequence itself and identified some characteristics that seemed to promote the kinds of mathematical activity we observed. These analyses point to an important interaction between the characteristics of the task sequence, the discourse in which the students engaged, and the opportunities for learning that resulted. This interaction is illustrated in Figure 3.

Our analysis suggests that specific aspects of the task promoted the kinds of discourse we saw as the students engaged with the initial tasks. For example, the students could approach the task by relying on their own intuition and aesthetic sense. This seemed to be a basis from which they felt comfortable providing justifications. Then, as the students shared their perspectives, they were exposed to other ways of thinking about symmetry (for example ways that included attention to rotational symmetry). This provided a chance for individual students to develop more expansive notions of symmetry which in turn set the stage for more sophisticated levels of discourse (for example the discussions about whether some rotations should be considered equivalent to each other.) Aspects of the follow-up task further promoted this shift by presenting students with two options for counting symmetries, resulting in mathematical argumentation to resolve

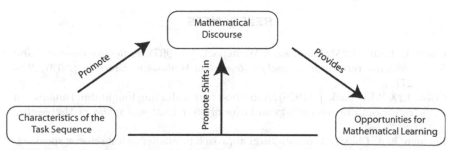

**FIGURE 3. Illustration of the interaction between tasks, mathematical discourse, and opportunities for mathematical learning.**

the question of whether (and, if so, under what conditions) two different motions should be considered to be equivalent symmetries. Ultimately this discourse supported the students' development of the concept of equivalence, which was an important mathematical goal of the course.

To summarize, we found that four characteristics of the Measuring Symmetry task sequence were particularly important in terms of promoting substantive mathematical discourse and as a result providing opportunities for mathematical learning:

- The tasks provided multiple access points.
- The tasks called for the creation of a mathematically significant model.
- The tasks admitted multiple *justifiable* approaches.
- The tasks provided a context that supported the resolution of the argumentation.

The first characteristic allowed all of the students to contribute to the mathematical discourse and engage with the task. The second characteristic ensured that the discourse would focus on mathematically significant ideas. The third characteristic set the stage for a substantive mathematical debate. Finally, the fourth characteristic supported the instructional goal (in this case the learning of equivalence) by providing a means to evaluate various approaches on the way to settling on the most mathematically viable approach.

Together these aspects of the Measuring Symmetry task seemed to support rich mathematical discourse and as a result provided opportunities for learning. Further research is needed to determine the extent to which these characteristics (or others) are necessary and/or sufficient. Nonetheless, this case of the Measuring Symmetry task sequence illustrates the important role that task design can play in promoting discourse and learning, and begins the process of identifying design principles for effective tasks.

## REFERENCES

Cobb, P., Boufi, A., McClain, K., & Whitenack, J. (1997). Reflective discourse and collective reflection. *Journal for Research in Mathematics Education, 28*(3), 258–277.

Cobb, P., & Whitenack, J. (1996). A method for conducting longitudinal analyses of classroom videorecordings and transcripts. *Educational Studies in Mathematics, 30*, 213–228.

Forman, E. A. (2003). A sociocultural approach to mathematical reform: speaking, inscribing, and doing mathematics within communities of practice. In Kilpatrick, J., Martin, W. G., & Schifter, D. (Eds.), *A research companion to principles and standards for school mathematics* (pp. 333–352). Reston, VA: National Council of Teachers of Mathematics.

Gravemeijer, K. (1999). How emergent models may foster the constitution of formal mathematics. *Mathematical Thinking and Learning, 1*(2), 155–177.

Harel, G. (2007). The DNR system as a conceptual framework for curriculum development and instruction. In Lesh, R., Kaput, J., & Hamilton, E. (Eds.), *Foundations for the future in mathematics education.* Mahwah, NJ: Erlbaum.

Hatano, G., & Inagaki, K. (1991). Sharing Cognition through collective comprehension activity. In Resnick, L. B., Levine, J. M., & Teasley, S. D. (Eds.), *Perspectives on socially shared cognition* (pp. 331–348). Washington, DC: American Psychological Association.

Hiebert, J. (2003). What research says about the NCTM standards. In Kilpatrick, J., Martin, W. G., & Schifter, D. (Eds.), *A research companion to principles and standards for school mathematics* (pp. 5–23). Reston, VA: National Council of Teachers of Mathematics.

Himmelberger, K. S., & Schwartz, D. L. (2007). It's a home run! Using mathematical discourse to support the learning of statistics. *Mathematics Teacher, 101*(4), 250–256.

Ilaria, D. (2002). *Questions that engage students in mathematical thinking.* Paper presented at the 24th annual meeting of the North American Chapter of the International Group for the Psychology of Mathematics.

Kazemi, E., & Stipek, D. (2001). Promoting conceptual thinking in four upper-elementary mathematics classrooms. *Elementary School Journal, 102*, 59–80.

Kysh, J., Thompson, A., & Vicinus, P. (2007). Welcome to the MT 2007 focus issue: Mathematical discourse. *Mathematics Teacher, 101*(4), 245–246.

Lampert, M. (1986). Teaching Multiplication. *Journal of Mathematical Behavior, 5*(3), 241–280.

Lampert, M., & Cobb, P. (2003). Communication and language. In Kilpatrick, J., Martin, W. G., & Schifter, D. (Eds.), *A research companion to principles and standards for school mathematics* (pp. 237–249). Reston, VA: National Council of Teachers of Mathematics.

Lesh, R., Cramer, K., Doerr, H., Post, T., & Zawojewski, J. (2003). Model development sequences. In Lesh, R., & Doerr, H. (Eds.), *Beyond constructivism* (pp. 35–58). Mahwah, NJ: Lawrence Erlbaum.

Lesh, R., & Lehrer, R. (2000). Iterative refinement cycles for videotape analyses of conceptual change. In Kelly, A. & Lesh, R. (Eds.), *Handbook of research design in mathematics and science education* (pp. 665–708). Mahwah, NJ: Erlbaum.

National Council of Teachers of Mathematics. (1991). *Professional standards for teaching mathematics.* Reston, VA: National Council of Teachers of Mathematics.

National Council of Teachers of Mathematics. (2000). *Principles and standards for school mathematics.* Reston, VA: Author.

Rasmussen, C., & Marrongelle, K. (2006). Pedagogical content tools: Integrating student reasoning and mathematics in instruction. *Journal for Research in Mathematics Education, 37*, 388–420.

Simon, M., & Blume, G. W. (1996). Justification in the mathematics classroom: A study of prospective elementary teachers. *Journal of Mathematical Behavior, 15*, 3–31.

Staples, M., & Colonis, M. M. (2007). Making the most of mathematical discussions. *Mathematics Teacher, 101*(4), 257–261.

Stein, C. C. (2007). Let's talk: Promoting mathematicnhWeaver, D., Dick, T., & Rigelman, N. (2005). *Assessing the quality and quantity of student discourse in mathematics classrooms.* Portland, OR: RMC Research Corporation.

# CHAPTER 6

# LEARNING TO USE STUDENTS' MATHEMATICAL THINKING TO ORCHESTRATE A CLASS DISCUSSION

**Blake E. Peterson and Keith R. Leatham**

Learning to orchestrate class discussions that are based on students' mathematical thinking is one of the most difficult aspects of learning to teach in ways that build on students' mathematical experiences. Based on a research project in which student teaching was restructured so as to focus on using student thinking, we describe the steps of a process that teachers move through when using students' mathematical thinking. We also identify some roadblocks that keep student teachers from listening to and understanding, from recognizing, and from effectively using student mathematical thinking for classroom discussion. We discuss how understanding this process and these roadblocks can be useful to mathematics teacher educators in their work with preservice mathematics teachers.

**KEYWORDS:** orchestrating classroom practice; preservice teacher education; secondary student teaching; student mathematical thinking; teacher knowledge; teaching practice.

---

*The Role of Mathematics Discourse in Producing Leaders of Discourse*, pages 99–128
**99**

Mathematics classrooms wherein the teacher promotes mathematical discussion based on students' mathematical thinking[1], and then orchestrates that discussion in ways that facilitate yet deeper mathematical thinking, provide opportunities for students to meaningfully "struggle with important mathematics," (Hiebert & Grouws, 2007, p. 387) something that research has shown is critical for students to learn with understanding (Hiebert & Grouws, 2007; Hiebert et al., 1997; National Council of Teachers of Mathematics, 1991, 2007). Orchestrating such discussions, however, seems to be one of the most difficult aspects of this approach to teaching (Sherin, 2002a), particularly for novice teachers. This paper reports on the results of a research project that put student teachers (STs) in situations where they were trying to elicit students' mathematical thinking, and then studied how they navigated the road to using that thinking in their teaching.

## USING STUDENTS' MATHEMATICAL THINKING

The NCTM (2007) recommended that mathematics teachers "orchestrate discourse by... listening carefully to students' ideas and deciding what to pursue in depth from among the ideas that students generate during a discussion" (p. 45). It is this careful listening to, and pursuit of students' ideas that we label as *using students' mathematical thinking*; we refer to such opportunities to use students' mathematical thinking as *teachable moments*. Although *teachable moment* is not a well-defined construct in the literature, the idea of teachers capitalizing on students' mathematical thinking "in the moment" is frequently discussed in the literature on mathematics classroom discourse (e.g., Doerr, 2006; Manouchehri & St. John, 2006; Schoenfeld, 2008). In this section we describe a conceptualization of the process of using students' mathematical thinking that both informed and evolved over the course of our study. In order to effectively use students' mathematical thinking, teachers need to:

1.  listen to and understand student thinking;
2.  recognize the thinking as a teachable moment; and
3.  use the thinking for a mathematical and pedagogical purpose.

### Listening and Understanding

Effective teaching "requires careful listening" (Erickson, 2003, p. ix) in large measure because effective teaching builds on what and how students

---

[1]By "students' mathematical thinking" we mean, among other things, their solution strategies, their justifications and reasoning, and their models and representations.

(particularly those present) think. In order to use students' mathematical thinking, teachers need to listen with the intent of using that thinking in order to build the classroom understanding of the mathematics at hand. Such listening must occur both when teachers explicitly elicit their students' ways of thinking as well as in the myriad moments that arise unexpectedly. Although careful listening creates teachable moments that serve meaningful purposes beyond content (Schultz, 2003), the listening to which we refer in this paper is content-specific. We are speaking of listening to students' *mathematical* thinking with the intent to use that thinking in order to further the learning of *mathematics* for all students in the classroom.

From our countless daily interactions, each of us can attest to the fact that it is possible to listen yet not understand. Thus, teachers may listen to students explain their thinking but may not understand that thinking. In order to understand students' mathematical thinking, teachers themselves must have an understanding of the mathematical concepts at hand (Ball, Lubienski, & Mewborn, 2001; Ma, 1999; Sherin, 2002b). Although we find it valuable to consider *listening* and *understanding* as separate steps in the process of using students' mathematical thinking, it is often difficult to distinguish between the two in practice.

Listening to students has long been both advocated and studied by educators (e.g., Confrey, 1993; Davis, 1997; Paley, 1986; Schultz, 2003). Davis, for example, considered different types of listening that led to different types of teacher actions. Listening was characterized as *evaluative* when it was somewhat superficial and as *interpretive* when it sought merely to understand; with both types of listening there was no apparent intention by the teacher that the results of listening and understanding students' thinking would inform or redirect the lesson. Such listening could be classified as *funneling* (Wood, 1998), wherein a teacher listens for student thinking that will lead toward a preconceived "best solution" and away from alternative and wrong strategies.

In contrast to evaluative and interpretive listening, Davis' (1997) *hermeneutic* listening had at its very core the notion that students' thinking would in large measure determine the direction of the lesson. Wood's (1998) *focusing* pattern encompasses Davis' (1997) hermeneutic listening as well as other types of pedagogically-sound listening. In the focusing pattern, the teacher listens for alternative and incorrect strategies as a means of elevating (and focusing) students' mathematical thinking toward important mathematical ideas. We thus adopt *focused* listening to describe the listening in our process—listening with the intent to understand and then to meaningfully use students' mathematical thinking in order to further mathematical objectives.

*Recognizing the Teachable Moment*

Once a teacher has listened to and understood a student's mathematical thinking, he must then *recognize* this moment as a teachable moment; he must recognize the potential pedagogical and mathematical value in pursuing that thinking in order to be able to eventually use it. Although good intentions and common content knowledge are likely sufficient to allow a teacher to listen to and understand students' thinking, recognizing such thinking as a teachable moment takes a great deal of specialized knowledge. Recognizing such moments may also depend on the goal of the lesson, unit, or course and may depend on how the student thinking that was shared fits with the flow of the lesson. Lewis & Tsuchida (1998) quoted a Japanese teacher as saying, "A lesson is like a swiftly flowing river; when you're teaching you must make judgments instantly" (p. 15). Recognizing shared student thinking as a teachable moment is one of the instantaneous judgments that are made during lessons. Recognizing the pedagogical and mathematical value in students' thinking in the moment is a difficult step in the process of using students' thinking, even for experienced teachers (Chamberlin, 2005).

*Using Students' Mathematical Thinking Effectively*

Once a teacher has listened in a focused way to students' mathematical thinking, has understood that thinking, and has recognized the mathematical and pedagogical value of that thinking, he is in a position to use that thinking. In our conceptualization, the purpose of such use is to help all students to gain a better understanding of the concept at hand. Thus the thinking that is used may be correct or incorrect and this use is more than just sharing different methods for solving the same problem. Effective use of students' mathematical thinking requires the teacher to orchestrate a discussion about the connections between the different methods or discuss why some methods work and others do not. Effective use involves more than explanations of the methods or thinking; it involves making explicit the reasoning and the mathematics behind the thinking.

*The Influence of Teacher Knowledge on the Process*

We have found Hill, Ball & Schilling's (2008) conceptualization of teachers' knowledge to be a useful way to think about our process conceptualization—how one listens to, understands, recognizes as valuable and uses students' mathematical thinking. In their conceptualization, Hill et al. consider subcategories of Shulman's (1986) content knowledge and pedagogical content knowledge. These subcategories begin to delineate some of the specialized knowledge that teachers have and that others do not. Thus, although mathematics teachers have mathematical knowledge shared with those who are not teachers (common content knowledge—CCK), teachers also have specialized mathematical knowledge; this knowledge, such as knowledge of the affordances and constraints of various mathematical representations and models, is not commonly held by the general public or even by working mathematicians, but is integral to the work of a mathematics teacher (specialized content knowledge—SCK). Hill et al. (2008) also consider different types of pedagogical content knowledge: knowledge of content and students (KCS), knowledge of content and teaching (KCT), and knowledge of curriculum.

We now consider how these types of knowledge aid us in viewing the process of using students' mathematical thinking. For example, in reflecting on our own experience as mathematics teachers, a common scenario often gives rise to teachable moments—students often make mathematical statements that are slightly incomplete or incorrect. Various types of mathematical knowledge for teaching help us to take advantage of such teachable moments. Our KCS might prompt us to listen very closely to the response to a particular question because we know that students often have misconceptions in this area and that an incomplete answer might be masking just such a misconception. Our knowledge of curriculum might help us to recognize that the incompleteness of this response is likely connected to a bit of mathematics that is just on the mathematical horizon (Ball, 1993), and thus one that would be valuable to pursue. Our SCK might help us use the incompleteness of this response as just the right motivation to discuss a different representation or model for the mathematics at hand. We thus view mathematical knowledge for teaching as critical for teachers to be able to carry out the process of effectively using students' mathematical thinking.

### Illustrating the Process Conceptualization

In order to illustrate the process of using students' thinking that we have just described, we present here an episode from our data wherein an ST uses her students' mathematical thinking quite effectively. We then point to the evidence within the episode and from the STs' reflections that allow us to use the *process conceptualization* in order to interpret the episode.

ST Emily[2] began her pre-algebra lesson by having students work individually for several minutes on three true-false questions regarding similarity

---

[2]All names for STs and students are pseudonyms.

of geometric figures. She then began a class discussion by asking for volunteers to share their decision on the first statement: "All squares are similar." Christopher volunteered and stated that he thought that the statement was true because, in a given square, all of the sides are equal, all of the angles are equal, and opposite sides are parallel. ST Emily followed up briefly and then asked whether anyone thought the statement was false. Brandon volunteered, but also argued that the statement was true, although for a different reason:

> Brandon: I think it's true because, no matter what, they both have parallel lines, and if you just draw two sets of parallel lines you get a square.
> ST Emily: Okay. What do parallel lines tell you about them being similar?
> Brandon: You have to have two sets of parallel lines to be a square, so obviously, if it's a square they all have parallel lines so that's what's similar.
> ST Emily: So, if something has parallel lines, that makes it similar?
> Brandon: Well, to be a square they have to have parallel lines. If they're a square, they all have parallel lines. So if you have like four squares, they're all similar because they all have two sets of parallel lines.

By this point in the conversation there were many students in class who were shaking their heads, saying "no, that's not right," and raising their hands in hopes of responding to Brandon's thinking.

> ST Emily: Okay. What do you think about that Kayla?
> Kayla: I think that's not true—it is true, but—.
> ST Emily: Okay. What's not true and what is true?

Kayla then went to the board, drew two squares (one with side length 1 unit and one with side length 2 units), and then argued that there was a common scale factor of 2 between the two squares. Although it is never stated explicitly, apparently Kayla agreed with Brandon about the pairs of parallel lines always existing in squares, but not in his use of this reasoning to conclude that all squares are similar. Samantha then asked Kayla why she included the scale factor as part of her explanation, to which Kayla responded, "Because that's what we found for similarity last time." Samantha then continued the conversation:

> Samantha: You could also kind of verify with a square and a rectangle. Rectangles, they have to have parallel sides, but squares, all their side lengths have to be the same.
> ST Emily: Okay. So how is that different from what Brandon was saying?

> Samantha: Brandon was saying that they just had to be parallel.
> ST Emily: So can you show us a rectangle that's not similar—that has parallel sides?

Samantha went to the board, drew a skinny rectangle and a square, and pointed out that each figure has pairs of parallel sides. Then ST Emily asked the class whether the square and the rectangle are similar:

> Terry: Yeah, they have parallel lines. Because that one on top is the same as that [apparently comparing the top and bottom sides of the square].
> ST Emily: Okay, so we know that they are parallel, but does that make them similar? [Various students say "No" and "Maybe."] What do we have to know in order for something to be similar? [Numerous students, including Brandon, start to share their responses, some mentioning scale factor.] Let's listen to Brandon for a second. What did you say?
> Brandon: Similar means they have something in common. The thing that they have in common is parallel sides. [Lots of murmuring amongst the students]
> ST Emily: It's true that when we're talking about something being similar they have to have something in common. But when we're talking about something being *mathematically* similar—
> Alex: It has to— You have to find the scale factor.
> Cody: Doesn't it have to have the same angles and the same sides?

ST Emily then orchestrated a discussion with Cody and others about the two criteria for figures to be mathematically similar (i.e., scale factor and equal angles). Having done this, ST Emily turned the conversation back to Brandon.

> ST Emily: So, Brandon, is it enough for the lines to be parallel for it to be similar?
> Brandon: No.
> ST Emily: No. Why not?
> Brandon: Because they have to have corresponding sides and then corresponding lengths and stuff like that. Just parallel sides wouldn't be enough.
> ST Emily: Okay. Do we all understand that? It's true that they do have something in common, Brandon, but they're not mathematically similar.

We now analyze this episode according to the process of using students' mathematical thinking. We find evidence that ST Emily listened to Brandon's initial contribution in her first follow-up question, which also represents ST Emily's attempt to better understand Brandon's thinking. Her second question demonstrates that she understood his claim. She later reflected on the beginning of this conversation in this way: "I continued to push him to explain until I understood that his definition of similar was having something in common." At least in part, it was ST Emily's KCS that contributed to her ability to listen to and understand Brandon's thinking.

That ST Emily recognized this situation as a teachable moment and felt as though she had taken advantage of it was also revealed in her reflection: "Then I was able to validate his thinking and talk about what it means to be mathematically similar." We find this statement to be somewhat understated. ST Emily allowed the class to respond to Brandon's thinking and that thinking did not always directly address the underlying issue Brandon had raised. ST Emily brought the conversation back to Brandon's thinking on several occasions, thus helping to focus Brandon and the rest of the class on the distinction between the common and the mathematical definitions of *similar*. Her SCK provided her the ability to draw this distinction and her KCT helped her to direct the class discussion in that direction. This episode thus demonstrates the process of listening to and understanding, recognizing, and effectively using students' mathematical thinking and the role that mathematical knowledge for teaching played in supporting ST Emily in carrying out that process.

## METHODOLOGY

This study took place in the context of a larger project wherein we altered the structure and purpose of student teaching in an attempt to emphasize the elicitation and use of students' mathematical thinking while deemphasizing survival and classroom management. In this student teaching project, a pair of STs was placed with one cooperating teacher and two or three pairs of STs at different schools formed a cluster. The STs taught at most one lesson per week during the first 5 weeks of their 15-week student teaching internship. These lessons were planned in their pairs but were taught individually and observed by the other STs in the cluster, the cooperating teacher and the university supervisor. Following the teaching of the lessons, a reflection meeting was held in which the STs who taught the lesson would reflect on the goals of their lesson and on how the tasks of the lesson were intended to meet those goals. The observers then had an opportunity to ask questions and to make comments. As part of the altered structure, the STs also conducted directed observations and student interviews, and wrote weekly reflection papers (all focused on students' mathematical thinking) as a means of processing and synthesizing what they were learning. For more details on this restructuring of student teaching see Leatham and Peterson (2009).

For this study six female STs (Emily, Christina, Holly, Megan, Jennifer, and Ashley) and three cooperating teachers were purposefully selected to participate. The STs were chosen based on feedback from their past professors, who were asked to recommend students who they felt were primed to excel during student teaching. Emily and Christina were placed in a middle school and Holly and Megan were placed in a junior high school. These four STs were placed with teachers who were approaching their instruction from an NCTM *Standards* perspective and were using a reform-based curriculum. Jennifer and Ashley were placed in a high school setting with a new teacher who taught fairly traditionally but was open to learning new ideas and who supported the STs in implementing such ideas.

The data for this study consisted of video recordings of all ST lessons and the accompanying reflection meetings as well as the reflection papers that the STs wrote regarding these observations and reflection meetings. To identify candidates for teachable moments, the reflection meetings were analyzed for specific comments made by the person who taught the lesson or by observers. The comments of interest were those that made reference to student thinking observed in the lesson. Once these comments were identified, the lesson was analyzed to locate the episode that was being referenced. The episodes were then analyzed to determine whether they were teachable moments. This determination was based on whether both researchers felt that they might have pursued the students' mathematical thinking had they been teaching the class. In addition, the reflection papers written by the STs who taught the lessons were analyzed to identify any additional thoughts they had on the identified episode.

Once these teachable moments, as well as any comments or reflections about them, were identified an analysis of how the teacher used the student thinking ensued. In that analysis we viewed the episode through the lens of our conceptualization of the process of using students' mathematical thinking, seeking evidence as to whether each step was accomplished by the ST. Having identified the stopping points in the process, we looked for evidence of why the ST stopped the process where she did. This assessment was done by evaluating comments made during the lesson, during the reflection meeting or in the reflection paper. From these various data sources, we identified and describe here a variety of roadblocks that inhibited these STs from effectively completing the process of using student mathematical thinking.

## ROADBLOCKS TO EFFECTIVE USE OF STUDENTS' MATHEMATICAL THINKING

All the STs believed to some extent that their lessons would be more productive if their students were given opportunities to make comments or to share their solutions to problems. Therefore, there were many times during their lessons when they elicited students' mathematical thinking; often these instances could be viewed as teachable moments. As was expected of novice teachers, and regardless of whether the STs were using reform curricula in a middle school or traditional curricula in a high school, they ran into similar issues when trying to conduct a whole class discussion that used their students' thinking in order to assist all students to come to a deeper understanding of the underlying mathematics. We describe here a collection of roadblocks that hindered the process of effectively using student thinking for classroom discussion. Although we use the term "roadblocks," we view such instances in a positive light. As mathematics teacher educators we were pleased to see our STs grappling with these important issues—bumping up against important dilemmas of teaching. The identification of these roadblocks informed us as to where we needed to go as teacher educators in our efforts to help the STs continue their development as mathematics teachers.

### Roadblocks to Listening and Understanding

The literature is replete with examples of novice teachers (e.g., Borko et al., 1992; Cooney, 1985; Schultz, 2003) and experienced teachers (e.g., Ball, 1993; Davis, 1997; Lampert, 1990; Schultz, 2003) who struggle to attend to the complexities of teaching. It is no small task for teachers to balance attending to what students are saying with attending to what they will do or say next. Thus, one major roadblock to listening seemed to be the inability to attend to student thinking *and* attend to other aspects of teaching. In addition, even when the STs were listening to the substance of their students' thinking, they sometimes struggled to *understand* what was being said. When leading a class discussion where students are encouraged to share their thinking and methods for solving a problem, a ST's knowledge or experience may not allow her to understand a student's thinking. Because the student strategy is unique or different, the ST may not understand the point the student is trying to make, even though she is listening. Student thinking that is not understood cannot be used to enrich the class discussion for the benefit of all students.

An example of a roadblock to listening and understanding occurred as ST Jennifer taught a pre-calculus lesson that she had planned with her partner ST Ashley. The task that they had created was a set of cards containing different linear functions represented using words, graphs or equations. The students were asked to classify the cards according to their attributes, such as increasing or decreasing slopes. The STs wanted the students to reflect on the attributes of parallel, perpendicular, vertical and horizontal lines by looking at the similarities and differences among the various representations of a line. However, the STs had created this task by adapting an activity they had done in a university class, where each of several different kinds of functions was represented in four different ways – numerically, graphically, verbally and algebraically (see Cooney, Brown, Dossey, Schrage, & Wittmann, 1996, pp. 41-45). One of the primary purposes of this original task was to help preservice teachers review many different types of functions while simultaneously considering the attributes of these multiple representations. The STs adapted the original task to have three different representations of each of several linear functions that varied according to their slope or y-intercept. Because this "matching" characteristic of the original task still existed, many of these pre-calculus students attempted to group their cards only according to the three different representations of the same function, rather than considering classifications based on more general characteristics such as increasing or decreasing slopes.

Although much of the expressed student thinking was focused on grouping the cards according to the different representations of the same function, some of the thinking was clearly related to slope, one of the intended foci of the activity. However, as ST Jennifer elicited her students' thinking as part of a class discussion, she simply said, "Okay, okay. Yeah, that's interesting." and then moved on. A similar behavior was seen as ST Jennifer moved from group to group prior to the class discussion. She asked one group how they were classifying their cards but did not ask any follow up questions about what they were looking for or why. After looking at another group's work, she said "Oh, [you are classifying] by y-intercept. Good job." This student thinking seemed to meet the lesson goal and yet she had no further discussion beyond this comment. Thus, ST Jennifer was not listening to her students' thinking, even though some of it could have helped her to meet her mathematical goals.

In the reflection meeting, ST Jennifer was asked to describe the classification she was hoping to see. She responded by saying:

> We didn't really expect them to say, "Okay, well, this is the graph, it matches this equation, it matches this story." We didn't think they'd do that right off the bat.... Every single group did that.... The first thing that they came up with every time is, "We just matched them up."

Throughout this discussion ST Jennifer continually returned to the problematic nature of the unanticipated responses. We believe that she was so preoccupied with the number of unanticipated solutions that she did not listen for the thinking that might lead to her goal. In this case, ST Jennifer's attention to the seeming failure of her task hindered her from listening to her students' thinking.

Concerns about classroom management also functioned as roadblocks to listening and understanding. With respect to this same card-sorting lesson, ST Jennifer said the following in her reflection paper:

> I was noticing that some students were finished with the activity, so I asked them to write their answers on the board to give them something to do. Had I been more aware of their answers, I would not have had them present. Their answers were not beneficial to the class discussion.

In this case, ST Jennifer did not listen to her students' thinking before she attempted to *use* it. Her attention to issues of classroom management hindered her from listening to the student thinking that she was observing in the class.

In summary, two of the roadblocks to listening and understanding are the challenge of keeping the classroom running smoothly when the lesson feels like it is falling apart and using student thinking for a management purpose instead of using it to better understand the mathematics. As mentioned previously, effective teaching is a complex endeavor. Learning to use students' mathematical thinking requires learning to attend to that thinking while attending to many other aspects of the class and of the lesson.

## Roadblocks to Recognizing the Teachable Moment

It is one thing to understand what a student says. It is quite another thing to recognize that thinking as a teachable moment—to understand the significance of what the student has said and to see value in that thinking from a mathematical and pedagogical stance. We have identified a number of roadblocks to such recognition.

### Assumption of Understanding

STs often work on the assumption that their students already understand the mathematics at hand. Our data revealed two variations of how these assumptions play out.

*Fill in the Blanks.* Novice teachers have a tendency to implicitly "fill in the blanks" when their students are talking about mathematics rather than asking the students to do so (cf. Ball, 2001). Students often use imprecise language when answering questions or sharing their work. The STs frequently forgave this imprecision, assuming that the student understood what they were superficially or inadequately describing. They failed to recognize such moments as important opportunities to push the student to clarify their statements and thinking.

An example of this roadblock occurred in a lesson taught by ST Jennifer. One of the homework questions had asked the students to find a line through the point (6, 5) that is perpendicular to:

$$y = -\frac{2}{3}x + 2\cdot$$

As a class the students had arrived at the equation:

$$y = \frac{3}{2}x + b\cdot$$

ST Jennifer then asked what they needed to do next for the line to pass through (6, 5). ST Jennifer described the student response and her thinking as follows:

> I got the chorus answer "you plug in (6,5)." I then assumed that most students knew how to do this and moved on. After the reflection meeting, I now see this situation differently. If I was in this situation again, I would ask why I can't plug in any value I want to. This discussion probably could have deepened students' understanding of an equation for a line. I hope that next time, I can be more aware of little situations like this that could strike up a mathematically engaging discussion.

It is clear from ST Jennifer's comments that she assumed the students understood the underlying mathematics when they said "you plug in (6, 5)" and filled in the blank about how and why plugging in (6, 5) yields the desired results.

*Simply Remind.* When students display incorrect or incomplete thinking about mathematics that has been recently talked about in class or that they have learned in the past or that was written in the instructions, the STs tended to assume that the students actually understand it and that they "just need to be reminded" about it (thus equating learning with memorizing). Such incomplete or incorrect student thinking was often viewed by the STs as "mistakes" rather than "misunderstandings." In these instances the STs tended not to question students' understanding of that mathematics. Instead they either reminded the students of the time or place where the concept had been addressed before or rephrased the student comments by correcting or completing the response.

An example of this roadblock occurred when ST Holly was teaching a lesson wherein the students were asked to complete a table describing the distance from a motion detector at time *t* as a person walks toward it. (Although this task was carried out without a motion detector, the students had been involved in an activity where they had used the motion detector earlier in the week.) The students were given tables with several values of $t$ already included and were asked to complete the tables and to graph their results. Some of these values of $t$ could be interpreted to mean that the person had walked past the motion detector. The table was labeled with time as the independent variable and distance as the dependent variable. The STs wanted the students to enter negative distances once the person walked past the motion detector even though this solution was not clear from the context. There were many students who did not use negative numbers in their solutions. As ST Holly interacted with those groups or students she gave a variety of little hints about how they might fill in the paper. In her reflection paper she commented that she had tried to resolve the issue that students were not using negative numbers in their solutions by encouraging "the students to read the problems carefully and to make sure what they were saying." She also said, "I know the students just didn't read the problem correctly." ST Holly did not see the situation as being problematic for the students and assumed that they would have understood if they had just read the instructions more carefully. Her tendency to view the students' alternate solutions as the result of mistakes rather than as attempts to make meaning of the task hindered ST Holly from recognizing this thinking as a teachable moment.

In each case here, the STs' assumption of understanding inhibited them from recognizing the moment as a teachable moment. They assumed that when a student provided a simple response that was correct, the student had the desired depth of understanding the STs were seeking. They also assumed that when a student made an incorrect statement, they had "just forgotten" but they really understood the concept at hand. Both of these types of assumptions kept the STs from recognizing the potential rich conversation that could occur if only they dug a little deeper.

### Funneling

The other main roadblock to recognition that we identified in our data was referred to earlier in this paper: "funneling" (Wood, 1998), or looking for a particular response and, in the process, failing to recognize the mathematical and pedagogical significance of other responses. It may seem as though funneling could be categorized earlier in the process, as a roadblock to listening, but we do not think this is the case. In order to funnel, one must actually listen to student thinking and understand it enough to recognize that it is not the thinking being sought. Thus, STs who funnel have at least listened and understood. What they fail to do is to recognize the mathematical and pedagogical significance of the response. Because they have a preconceived notion of the response that will lead to a teachable moment, they fail to recognize other responses that may lead to similar or even different but still valuable teachable moments.

An example of funneling that prevented a ST from recognizing a teachable moment occurred when ST Christina was launching a lesson wherein students were going to input equations into a calculator and look at the tabular outputs to make decisions about the situation. In anticipation of this approach, ST Christina asked the students to identify the independent and dependent variables in the equations $A = \pi r^2$ and $C = 2\pi r$. This activity was meant to be a quick review so that the students would be able to input equations properly into the calculator. ST Christina wanted to hear that $r$ was the independent variable and that $A$ and $C$ were the dependent variables in their respective equations. After a student had shared his response that $r$ was the independent variable and $A$ was the dependent variable in the former equation, Morgan said that she thought $C$ was the independent variable and $r$ was the dependent variable in the latter equation. Because this response was not what ST Christina wanted, she began a funneling process:

> ST Christina: Morgan, come up and explain to us what you have here.
>
> Morgan: I did the circumference because the radius depends on how big or small the circle is. So I said the circumference is independent and the radius is dependent on the circumference.
>
> ST Christina: OK. Thanks Morgan. Who has the same thing as Morgan? Who has something different? Who doesn't know? [pause] Who said they have something different? Brian's the only one who has something different?
>
> Sage: I don't know.
>
> ST Christina: Nobody else? Katherine, do you? Do you want to explain?
>
> Katherine: I just said the independent would be the radius and the dependent would be the circumference.
>
> ST Christina: OK, Why?
>
> Katherine: Because... the circumference is the—. Wait, no, I agree with her [Morgan].
>
> ST Christina: Are you sure? You were going good there. Do you want to keep explaining what you were saying?
>
> Katherine: I was going to say that the circumference would change if the circle gets smaller. But um, you can find the circumference without the radius I think.
>
> ST Christina: You can find the circumference without the radius? How would you do that?
>
> Katherine: Um. I don't know.
>
> . . .
>
> ST Christina: Brian, what do you think?

Brian: Couldn't kind of both of them go both ways? Because like in area. Like as the area gets smaller so does the—. Oh, never mind.

ST Christina: So let's look back at this one. How did these equations relate with each other with the independent and dependent, um, with both of them and how can we think this through? Anybody have some ideas besides Brian? Brian, thanks for your help though. Abe, what do you think?

In this episode Morgan presented a solution that was not what ST Christina had anticipated, so she asked the class if someone had approached the situation differently and this is where the funneling began. Katherine started to say that she disagreed with the first student and then changed her mind. ST Christina tried to pursue Katherine's initial thinking because it was what she was looking for, namely $r$ as the independent variable and $C$ as the dependent variable. Brian then suggested that it could go either way and then backed off as did Katherine. In this case, however, the ST Christina did not pursue Brian's comment. With both Katherine and Brian ST Christina funneled toward her preconceived correct solution and away from the divergent thinking of the students. This funneling had ST Christina trying to get Katherine to explain her original thoughts because they supported her goal and yet the funneling hindered ST Christina from recognizing the richness of Brian's thinking as a teachable moment.

## Roadblocks to Effectively Using Students' Mathematical Thinking

Because STs are in the process of learning how to teach, it comes as no surprise that they might listen to student thinking, understand what students have said, recognize that thinking as a teachable moment, yet still not be able to use the student thinking in a way that furthers their mathematical learning goals—that capitalizes on the teachable moment. Our analysis of the data revealed a number of roadblocks to effective use of student thinking. In our categorization of these roadblocks, we make a distinction between roadblocks to trying to use student thinking and roadblocks to effective use. The former roadblocks inhibited the STs from even attempting to use their students' thinking. With the latter roadblocks, the STs attempted to use their students' thinking but fell short.

### No Attempt to Use

First, we located a number of instances in our data where the STs were able to recognize student thinking as a teachable moment and yet they made no attempt to use that thinking in their lesson (i.e., they did not pursue the thinking with the class). When such instances occurred we could

**Price Customers Would Pay**

| Total Price | Number of Customers |
|---|---|
| $150 | 76 |
| $200 | 74 |
| $250 | 71 |
| $300 | 65 |
| $350 | 59 |
| $400 | 49 |
| $450 | 38 |
| $500 | 26 |
| $550 | 14 |
| $600 | 0 |

**FIGURE 1. Table from Connected Mathematics 2 (Lappan et al., 2006, p. 32)**

trace the reason to a lack of knowledge, usually a lack of either SCK, PK, or curricular knowledge. We use several episodes to illustrate such roadblocks:

*Lack of SCK.* ST Holly had given her pre-algebra class the table shown in Figure 1, which gives the number of people out of 100 surveyed who would go on a bike tour for the given total prices. The students were asked the following question: "To make a graph of these data, which variable would you put on the *x*-axis? Which variable would you put on the *y*-axis? Explain." The students were also asked to "make a coordinate graph of the data on the grid paper" (Lappan, Fey, Fitzgerald, Friel, & Phillips, 2006, p. 32).

ST Holly had anticipated "that students would have problems with the independent and dependent variables," so during the first few minutes of the task she "went around to the different groups... looking for students that had their independent and dependent variables labeled correctly and also students that had their variables mixed up." Having noticed that many students seemed to be struggling a great deal with the identification of the independent and dependent variables, ST Holly decided to bring the class together for a class discussion. In order to initiate the class discussion she polled the class:

ST Holly: How many of you guys think that the *total price* is the independent? [Quite a few students raise their hands.] How many of you think that the *number of customers* is the independent? [Some (but fewer) students raise their hands.] Okay. Someone who thinks that the *number of customers* is independent, will you tell me why?

Spencer: I'd say that it's the *number of customers* because the customers depend on their opinion of what the price should be.

ST Holly: Okay. So the number of customers depends on what the price will be?

Spencer: No. The money depends on what the number of customers should be.

ST Holly: Okay. Is that what it says in the prompt? That's the one I did. Is there another idea why that one would be the independent one?

Breanna: You can't, because it's how many people want the price.

ST Holly: How many people want that price?

Breanna: Yeah. So, like, it's kind of hard to explain.

ST Holly: It's kind of hard to explain? Okay, someone who picked the *total price* to be the independent one, do you want to give me an explanation?

George: Yeah. I think it's the *total price* because, like, just because they want to go doesn't mean they can change the price, so the price stays the same. [ST Holly: Uhuh.]. So the number of people changes depending on if they want to pay that much or not.

ST Holly: Okay. Does everyone understand what he said? A little bit? Okay, let me give you another example.

ST Holly then proceeded to give an example that she and ST Megan had developed while planning their lesson, one that they had hoped would help students gain a better understanding of how the analysis of the context of a situation helps you to identify the independent and the dependent variables:

ST Holly: Ariel, let's say I come up to you and I'm like, "Do you want to buy my ipod?" What do you say?

Ariel: Um, no.

ST Holly: No? Why not? Why don't you want to buy my ipod?

Ariel: Well, probably because you may have used it already and everything and you already have your own songs that you already had on it.

ST Holly: Uhuh. That's a good thing. So it kind of depends on different things, right?

Ariel: Yeah.

ST Holly: So I can't just, like come up to you and be like, "Do you want to go on the bike tour?" You probably want to know how much it is first. Right?

Ariel: Uhuh.

ST Holly: So that's something to think about.

ST Holly then asked the students to get into pairs and return to the task.

In reflecting back on this class discussion ST Holly recognized that the discussion "never came to closure" on the issue of determining the independent and dependent variables. She then shared this important bit of insight into her thinking at the time:

> People would argue both sides and I could see both sides, but I didn't know how to justify them. And I just kept trying to bring it back to the context. But every time I'd bring it back to the context they'd come up with something different and I was like, "Oh, I didn't think of that." And so I really didn't know how to—. So I don't know.

In this episode ST Holly asked questions that allowed her students to share their thinking about determining the independent and dependent variables in this situation. There is evidence that she was listening to and understanding what they said (i.e., "I could see both sides"). She also recognized this discussion as a great opportunity to talk about the very thing she wanted to talk about—deciding which variable should be independent and which should be dependent. What seemed to keep ST Holly from using the students' thinking in this situation was a lack of SCK. ST Holly had developed a strong enough knowledge of independent and dependent variables to know that such classification was highly dependent on the context. She did not have sufficient understanding of the underlying mathematics, however, to allow her to see how to justify or refute the various responses that she received. Without this knowledge she was unable to push on that thinking, to point out what was correct and what was incorrect in the students' thinking. Rather than use the student thinking that had been proffered, her lack of SCK forced her to retreat from that thinking and to introduce her own thinking. In this episode, ST Holly's lack of SCK was a roadblock to her using her students' mathematical thinking.

*Lack of Pedagogical Knowledge.* Lack of pedagogical knowledge (PK) also impeded the STs from using their students' mathematical thinking. One of the most common ways this lack of knowledge revealed itself was when the STs would notice a productive conversation within a small group. The STs would often listen to such conversations, recognize them as valuable, and have the desire to use that thinking in a class discussion. They soon learned, however, that recreating such individual group discussions was not easy.

Their approach usually took the form of asking the group to share with the class the conversation they had just had; this approach rarely succeeded. As Ashley put it, "It was really fun to sit in on their argument, but for me, it's hard to recreate that argument in the classroom discussion because they feel like they've already had the conversation; they don't want to have the conversation again." In our experience, although students may indeed be reluctant to reproduce such conversations, such reproduction is practically impossible for them. Students tend to focus on the results of their conversations, seldom on the process or on pitfalls of those conversations. Thus, successful reproduction of valuable small-group conversations must be facilitated by the teacher, and usually entails involving the rest of the class in the issue that was being discussed, thus recreating the situation for the whole class that caused the productive conversation in the small group. This pedagogical knowledge of how to use students' mathematical thinking was not yet available to the STs in our study, and the lack of this knowledge was a roadblock to their use of that thinking. We note, however, the important progress the STs were making in that they were beginning to recognize their lack of this knowledge and to find ways to acquire it.

*Lack of Curricular Knowledge.* The STs' lack of curricular knowledge often impeded their ability to use students' mathematical thinking. For example, in the motion-detector task described earlier, many students questioned the notion of negative numbers in the context of the problem, because "the students were unsure about the motion detector being able to read the person if they were behind the motion detector" (ST Holly). Students who focused on the context of the problem questioned the use of negative numbers and many started to develop solutions that were building toward the notion of absolute value. The STs recognized that the solutions were headed in that direction and chose not to pursue them. As ST Holly stated, "I stayed away from that idea because I didn't want to talk about absolute value functions. I didn't even know if it was OK to talk about absolute value functions this early." In this situation ST Holly's lack of curricular knowledge impeded her from using the student thinking that was elicited by this task. She did not have a sufficient understanding of the connections between the current day's mathematics topic and the underlying mathematical ideas of absolute value.

Thus, in general, the STs in this study were often inhibited from using student thinking because they did not understand the mathematics enough to pursue the thinking with their students (lack of SCK), they did not know how to carry out that pursuit (lack of PK), or they did not know whether they "should" pursue it (lack of curricular knowledge). In each case the STs had sufficient knowledge to listen to, understand and recognize the value in their students' thinking; what they they lacked was the knowledge to use that thinking.

### Naïve Use

The STs who participated in this study often tried to use their students' mathematical thinking. This use, however, was not always effective. In analyzing their attempts to use students' mathematical thinking, we found a number of times when it was clear that the STs believed that they were indeed using the students' thinking effectively, although from our observation this was not the case. We classified such usage as naïve use—the STs were "technically" using their students' thinking, but such use was based on a naïve assumption about how students learn and did not really capitalize on the mathematical thinking of the students. The following sections describe the various types of naïve use that emerged from our analysis of the data.

*Student Thinking as a Trigger.* Using students' mathematical thinking as a trigger is somewhat akin to funneling. With funneling, however, the STs basically pass by all student thinking until they hear the thinking they are looking for—they then pursue or validate that thinking but fail to recognize the value in the other mathematical thinking that was shared. In the case of a trigger, the STs did recognize some value in what students' said, but the value is that they see a way that they can take some portion of what the students said (often not necessarily related to what the student meant) in order to redirect the conversation toward where they intended it to go. In terms of Woods' (1998) funneling and focusing constructs, when STs use students' mathematical thinking as a trigger they funnel but they *think* they are focusing—they think they are effectively using their students' thinking.

An example of using student thinking as a trigger comes again from the motion detector task and revolves around the issue of whether it was okay in this situation for the output numbers (distances) to be negative. In ST Megan's class she invited several students to put their answers on the board. Marcus then explained how they found the values in their table:

> Marcus: He went to six, one, and then—it would usually be negative four, but it didn't say it was in front or behind. So we just thought it was four feet away from the [motion detector] because it went back up again.
>
> ST Megan: Okay. Was that confusing to anyone else, whether or not you could go into the negatives? I saw a couple of papers where there was some argument. Let's talk about this idea of whether or not you can go into the negatives.

It appeared that ST Megan has listened to the student thinking that had been presented and that she recognized this thinking as worth pursuing. Rather than pursuing the reasoning of this pair of students, however, ST Megan chose to use their explanation as a trigger to talk about why it *does* make sense to use negative numbers in this situation. The students' explanation contained their reasoning for using positive numbers, not negative numbers. In fact, the mathematics of their explanation is focused on arguing that the context of the problem calls for the use of positive numbers. ST Megan viewed this as an incorrect answer and chose to use the statement "it would usually be negative four, but it didn't say it was in front or behind" as a trigger to first focus on the difficulty in deciding and then to focus on putting forth arguments that the values should be negative. Although the student's explanation was technically used in this situation, the mathematics of that explanation was neither used nor valued. Instead, a phrase about the problematic nature of the decision was taken up and used in order to redirect the focus of the discussion. We categorize such use of students' mathematical thinking as a trigger as naïve use. ST Megan seems to believe that she is indeed using the students' thinking, but her use is for her own purpose, which in this case is actually to try to make a point that is opposed the point the student was trying to make.

## Mere Presence of the Correct Solution

It was fairly common for the STs to elicit students' mathematical thinking and then fail to do anything with that thinking. In some cases, we concluded that the student thinking was merely elicited, but never listened to, let alone recognized as valuable and then used. In other instances, however, analysis revealed a variation on this phenomenon that we felt clearly should be classified as using (although naïvely) student thinking: the STs clearly believed that they were using the student thinking that had been elicited, although this use was at best implicit.

An episode from ST Megan's classroom illustrates this naïve use of student thinking. ST Megan had engaged her students in the Bicycle tour task (see Figure 1). For this part of the task, the students were asked to respond to the question, "Based on your graph, what price do you think the tour operators should charge? Explain" (Lappan et al., 2006, p. 32). After the students had worked on the task for some time, ST Megan initiated a class discussion by asking students to share their answers. One pair of students said they should charge $150 because it was at that price that the most customers had indicated they would participate. ST Megan then asked for others to share their solutions and a pair of students volunteered $350 (the correct solution) and explained that they used a revenue table to come up with their solution. A brief discussion followed about this latter answer, in which ST Megan implied that this latter answer was correct, and then the class moved on.

In this episode ST Megan elicited two different solutions with different solution methods and justifications. The first solution was incorrect; the second was correct. We classify this situation as a teachable moment because there are two reasonable solutions on the board and the class is primed to make arguments about their validity. Such a conversation would bring up the important mathematical ideas of revenue and maximization. ST Megan listened to the students' solutions and ensured that she understood them. Now, it is tempting to characterize ST Megan as not having recognized the teachable moment, and in terms of the explicit discussion and comparison of the two solutions, that might be correct. However, we believe that ST Megan actually thought that she was using both students' thinking. She facilitated the presentation of both solutions and she implicitly endorsed the latter (correct) response. We believe that ST Megan was operating under the following naïve assumption regarding student learning: the presence of the correct solution clears up the misconceptions underlying the incorrect solutions. It is by looking at ST Megan's teaching here through the lens of this assumption that we classify it as naïve use. Such naïve use of students' mathematical thinking seems to be based in a lack of KCS. This lack of knowledge allowed ST Megan to use here students' mathematical thinking, but only naively, thus hindering her from effectively (in this case, explicitly) using that thinking.

*Mere Presentation of Multiple Solutions.* We briefly highlight a different but related naïve use of students' mathematical thinking that occurred fairly often in the STs' lessons. Sometimes the STs managed to elicit significant student thinking, had students record at the board and explain this thinking, and then the STs moved on. The goal of the lesson seemed to devolve into "student sharing," rather than developing mathematics from what the students were sharing, a phenomenon that has been previously noted in the literature (e.g., Ball, 2001; Silver, Ghousseini, Gosen, Charalambous, & Font Strawhun, 2005). Again, like before when the presence of the correct solution was interpreted as having cleared up the misconceptions underlying incorrect solutions, in these situations the STs seemed to be operating under the assumption that the connections between the multiple presented solutions, and the mathematics that could be derived from exploring those connections were evident to the students—that the students had learned important mathematics simply from being exposed to multiple solutions or solution strategies. Again it is under this assumption that we classify such an approach as naïve use of student thinking.

*Incomplete Use*

As we have mentioned previously, the STs often did manage to use the student thinking that they listened to and recognized as valuable. The extent to which they were able to use it effectively, however, even when they tried, was often limited by their limited knowledge. Consider the following episode:

The students in ST Megan's class were considering some data that reported the amount of time certain students spent watching TV and those students' GPA. ST Megan asked the class whether *TV Time* or *GPA* would be the independent variable. One student answered *TV Time* and ST Megan follows up:

> ST Megan: What helped you decide it was *TV Time?*
> Jamie: Because it was related to time.
> ST Megan: Because it's related to time. [She notices another student with their hand raised.] Yes?
> Nate: Time usually ends up as the independent variable and so it should go on the *x*-axis.

ST Megan recognized this student thinking as worth pursuing, as she had noticed that her students appeared to be choosing *time* as the independent variable almost automatically if it showed up as one of the variables. This recognition is evidenced by what ST Megan did next:

> ST Megan: In some cases could it end up being the dependent variable? Do we have to be careful with that sometimes? What makes it hard to tell in this circumstance if its going to be independent or dependent? [pointing to a student who looks eager to contribute] Did you have an idea?
> Jenna: Well, because there could be different situations where they could both be either independent or dependent.
> ST Megan: Exactly. So, what kind of situation would we be thinking of if we said that *time* was dependent?

Taken together, Nate and Jenna's comments provided excellent student thinking on which ST Megan built toward a nice question that pushed her students to think more deeply about the mathematics of this situation. In particular, the class was poised for a discussion on determining the dependent and independent variables based on context. In the end, however, this conversation did not lead anywhere. The students were not able to think of a situation where *time* could be considered the dependent variable; most critically, neither was ST Megan able to construct such a situation. Because ST Megan lacked the SCK that would allow her to create such examples, she was not able to use effectively the student thinking in order to move the mathematical conversation towards her big mathematical idea—that the independent and dependent variables depend on the context (not just on whether *time* is one of the variables). ST Megan definitely recognized the shared student thinking as an opportunity to pursue important mathematical ideas, and she made a noble effort to do so. Her lack of SCK, however, hindered her ability to do so effectively and resulted in incomplete use of that thinking.

## DISCUSSION AND CONCLUSION

Our analysis of the data revealed numerous roadblocks to the steps in the process of using students' mathematical thinking. By and large, these road-blocks can be characterized by a lack of teacher knowledge; the STs often lacked the SCK, PK, KCS, or curricular knowledge to recognize or capitalize on teachable moments. At times, the STs recognized the value of their student's mathematical thinking and either did not pursue it or struggled to use it productively when they did pursue it because they did not have adequate knowledge of the representations or connections that would allow them to do so. At other times the STs heard and recognized student mathematical thinking that could be used to help all students better understand the topic at hand but did not have the PK to have a productive discussion about that thinking (like in the case of how to use three incorrect solutions in order to get at important mathematics).

Lack of curricular knowledge also often inhibited the STs from using student thinking. We hypothesize that the further removed a mathematical concept is from the lesson at hand, the less likely it is that STs will have the curricular knowledge to capitalize on a teachable moment that gets at that concept. The curriculum can be viewed as a set of concentric circles (see Figure 2), where the topic of the day is in the center. The goals of the unit, the course and mathematics in general are related to and often underlie the day's goal but often feel far removed to STs. STs' limited view of the curriculum (i.e., knowledge of curriculum) thus inhibits them from pursuing worthwhile mathematics.

In many of the roadblocks to effective use we have described in this paper, the STs recognized the teachable moment but their lack of knowledge (SCK, pedagogical, curricular) prevented them from using student thinking or only allowed minimal use. We identified a number of roadblocks, however, to the recognition of teachable moments. It seems as though KCS, in particular a knowledge of how students think about and learn mathematics, was the primary type of knowledge that inhibited this recognition. A common roadblock to recognition was an assumption of understanding,

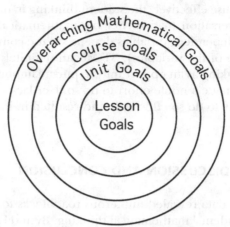

**FIGURE 2. A representation of the broadening layers of the mathematics curriculum.**

which comes from a lack of KCS. The STs' knowledge of students and how they learn led them to believe that once a concept had been "covered" (either by them or in a previous class) the students knew and understood that concept. If students seemed shaky with that knowledge, the STs tended to assume that the students had "just forgotten" what they had learned.

Another place where KCS had an influence on the productive use of student thinking was the naïve use. When the STs used the mere presentation of a correct solution as a way to clarify the flawed thinking that led to an incorrect solution, they exhibited their limited KCS. In this case, their KCS led them to believe that a student who had incorrectly solved a problem could resolve their misunderstanding by simply observing a correct solution, without having an explicit conversation about the approach. Similarly, the STs' KCS drove their approach of merely having the students share multiple solutions without any discussions of the connections between those solutions. Because of their lack of knowledge about how students learn, the STs assumed that the students would be able to see the similarities and differences between the different solutions without an explicit conversation about them.

It is interesting, however, to contrast this confidence in students' previous learning or their ability to make connections with the surprise (and doubt) that STs often express at the kinds of problems their students are able to solve. Thus STs tend to have high confidence in their students' previous learning abilities, but low confidence in their current learning abilities. Deeper KCS would likely foster quite the opposite set of assumptions about student learning, namely, a confidence in students' abilities to learn mathematics through solving problems on their own, but a healthy skepticism of their current mathematical understanding. Such skeptical optimism fosters an approach to teach wherein the teacher is very inquisitive about their students' thinking, always seeking to push on that thinking in the belief that such pushing will lead to great strides in student understanding.

The results of this study have important implications for mathematics teacher educators. First, our data demonstrate that STs are capable of learning to teach through focusing on their students' mathematical thinking. The revised student teaching structure in which these STs participated supported and encouraged their efforts to both elicit and use their students' thinking. These results add to a growing body of literature (e.g., Feiman-Nemser, 2001; Sowder, 2007) that refutes the logical fallacy that because novice teachers tend to begin with somewhat self-centered concerns, that teacher education programs should explicitly focus on addressing those concerns (Fuller, 1969). Our research demonstrates that STs are capable of focusing on and learning from their students' mathematical thinking.

Second, STs could benefit a great deal from an understanding of the steps of the process of effectively using student mathematical thinking:

1.  listen and understand student thinking;
2.  recognize the thinking as a teachable moment; and
3.  use the thinking for a mathematical and pedagogical purpose.

Although this process is certainly not the only way to teach in a way that is responsive to students' needs and thinking, the process does represent a tangible learning objective for novice teachers. Once the overall process is understood, STs are better able to reflect on their own teaching and that of others. They can focus on the points at which the process of using student thinking breaks down and on the type of knowledge that might help them to move further along in the process in the future. Also, as novice teachers better understand and value the process of using their students' mathematical thinking, they will realize that they need to plan into their lessons the time needed to pursue that thinking. It is difficult to discuss the mathematics of teachable moments if the time has not been allotted to do so.

Finally, teacher educators need to evaluate the degree to which their teacher education programs are designed to help novice teachers gain the knowledge needed to overcome these roadblocks. The content and structure of mathematics teacher education programs either afford or constrain the construction of this knowledge. Learning to teach activities, including open discussions with novice teachers about the pitfalls of an assumption of understanding, could help them to begin to develop the skeptical optimism needed to recognize and take advantage of teachable moments. These discussions with novice teachers about how students learn mathematics could also help them see the importance of making connections between solution strategies explicit through class discussions. Further activities designed to strengthen mathematical knowledge for teaching are then needed for novice teachers to develop the knowledge necessary to use those moments effectively.

Although much has been said about the importance of using significant mathematical tasks in order to elicit students' mathematical thinking, relatively little is known about the factors involved in using that mathematical thinking effectively. These results illustrate the complexity of this issue. More work needs to be done on designing and researching the effectiveness of "learning to teach" activities that can help novice teachers learn how to listen to, understand and recognize the value of their students' thinking, and then be able to use that thinking in order to orchestrate meaningful mathematical discussions. A good first step in this direction would be to discuss the process conceptualization and roadblocks presented here with preservice teachers as part of their teacher preparation program.

## REFERENCES

Ball, D. L. (1993). With an eye on the mathematical horizon: Dilemmas of teaching elementary school mathematics. *The Elementary School Journal, 93*, 373–397.

Ball, D. L. (2001). Teaching with respect to mathematics and students. In Wood, T. L., Nelson, B. S., & Warfield, J. (Eds.), *Beyond classical pedgogy: Teaching elementary school mathematics* (pp. 11–22). Mahwah, NJ: Lawrence Erlbaum Associates.

Ball, D. L., Lubienski, S. T., & Mewborn, D. S. (2001). Research on teaching mathematics: The unsolved problem of teachers' mathematical knowledge. In Richarrdson, V. (Ed.), *Handbook of research on teaching* (4th ed., pp. 433–456). New York: Macmillan.

Borko, H., Eisenhart, M., Brown, C. A., Underhill, R. G., Jones, D., & Agard, P. C. (1992). Learning to teach hard mathematics: Do novice teachers and their instructors give up too easily? *Journal for Research in Mathematics Education, 23*, 194–222.

Chamberlin, M. T. (2005). Teachers' discussions of students' thinking: Meeting the challenge of attending to students' thinking. *Journal of Mathematics Teacher Education, 8*, 141–170.

Confrey, J. (1993). Learning to see children's mathematics: Crucial challenges in constructivist reform. In Tobin, K. (Ed.), *The practice of constructivism in science education* (pp. 299–321). Hillsdale, NJ: Lawrence Erlbaum Associates.

Cooney, T. J. (1985). A beginning teacher's view of problem solving. *Journal for Research in Mathematics Education, 16*, 324–336.

Cooney, T. J., Brown, S. I., Dossey, J. A., Schrage, G., & Wittmann, E. C. (1996). *Mathematics, pedagogy, and secondary teacher education.* Portsmouth, NH: Heinemann.

Davis, B. (1997). Listening for differences: An evolving conception of mathematics teaching. *Journal for Research in Mathematics Education, 28*, 355–376.

Doerr, H. M. (2006). Teachers' ways of listening and responding to students' emerging mathematical models. *ZDM: The International Journal on Mathematics Education, 38*, 255–268.

Erickson, F. (2003). Foreword. In Schultz, K., *Listening: A framework for teaching across differences* (pp. ix–xv). New York: Teachers College Press.

Feiman-Nemser, S. (2001). Helping novices learn to teach: Lessons from an exemplary support teacher. *Journal of Teacher Education, 52*, 17–30.

Fuller, F. (1969). Concerns of teachers: A developmental conceptualization. *American Educational Research Journal, 6*, 207–226.

Hiebert, J., & Grouws, D. A. (2007). The effects of classroom mathematics teaching on students' learning. In Lester, Jr., F. K. (Ed.), *Second handbook of research on mathematics teaching and learning* (Vol. 1, pp. 371–404). Charlotte, NC: Information Age Publishing.

Hiebert, J., Carpenter, T., Fennema, E., Fuson, K. C., Wearne, D., Murray, H., et al. (1997). *Making sense: Teaching and learning mathematics with understanding*. Portsmouth, NH: Heinemann.

Hill, H. C., Ball, D. L., & Schilling, S. G. (2008). Unpacking pedagogical content knowledge: Conceptualizing and measuring teachers' topic-specific knowledge of students. *Journal for Research in Mathematics Education, 39*, 372–400.

Lampert, M. (1990). When the problem is not the question and the solution is not the answer: Mathematical knowing and teaching. *American Educational Research Journal, 27*, 29–63.

Lappan, G., Fey, J. T., Fitzgerald, W. M., Friel, S. N., & Phillips, E. D. (2006). *Connected mathematics 2, grade 8: Variables and patterns*. Upper Saddle River, NJ: Prentice Hall.

Leatham, K. R., & Peterson, B. E. (2009). Purposefully designing student teaching to focus on students' mathematical thinking. Manuscript submitted for publication.

Lewis, C. C., & Tsuchida, I. (1998). A lesson is like a swiftly flowing river: How research lessons improve Japanese education. *American Educator, 22*(4), 12–17, 50–52.

Ma, L. (1999). *Knowing and teaching elementary mathematics: Teachers' understanding of fundamental mathematics in China and the United States*. Mahwah, NJ: Lawrence Erlbaum Associates.

Manouchehri, A., & St. John, D. (2006). From classroom discussions to group discourse. *Mathematics Teacher, 99*, 544–551.

National Council of Teachers of Mathematics. (1991). *Professional standards for teaching mathematics*. Reston, VA: Author.

National Council of Teachers of Mathematics. (2007). *Mathematics teaching today: Improving practice, improving student learning* (2nd ed.). Reston, VA: Author.

Paley, V. G. (1986). On listening to what the children say. *Harvard Educational Review, 56*, 122–131.

Schoenfeld, A. H. (2008). On modeling teachers' in-the-moment decision making. In A. H. Schoenfeld (Ed.), *A study of teaching: Multiple lenses, multiple views, JRME monograph #14* (pp. 45–96). Reston, VA: National Council of Teachers of Mathematics.

Schultz, K. (2003). *Listening: A framework for teaching across differences.* New York: Teachers College Press.

Sherin, M. G. (2002a). A balancing act: Developing a discourse community in a mathematics classroom. *Journal of Mathematics Teacher Education, 5,* 205–233.

Sherin, M. G. (2002b). When teaching becomes learning. *Cognition and Instruction, 20,* 119–150.

Shulman, L. S. (1986). Those who understand: Knowledge growth in teaching. *Educational Researcher, 15*(2), 4–14.

Silver, E. A., Ghousseini, H., Gosen, D., Charalambous, C., & Font Strawhun, B. T. (2005). Moving from rhetoric to praxis: Issues faced by teachers in having students consider multiple solutions for problems in the mathematics classroom. *Journal of Mathematical Behavior, 24,* 287–301.

Sowder, J. T. (2007). The mathematical education and development of teachers. In Lester, Jr., F. K. (Ed.), *Second handbook of research on mathematics teaching and learning* (Vol. 1, pp. 157–223). Charlotte, NC: Information Age Publishing.

Wood, T. (1998). Alternative patterns of communication in mathematics classes: Funneling or focusing? In Steinbring, H., Bartolini Bussi, M. G., & Sierpinska, A. (Eds.), *Language and communication in the mathematics classroom* (pp. 167–178). Reston, VA: National Council of Teachers of Mathematics.

# CHAPTER 7

# ORCHESTRATING WHOLE-GROUP DISCOURSE TO MEDIATE MATHEMATICAL MEANING

**Mary P. Truxaw and Thomas C. DeFranco**

The most common pattern of classroom discourse follows a three-part exchange of teacher initiation, student response, and teacher evaluation or follow-up (IRE/IRF) (Cazden, 1988/2001). Although sometimes described as encouraging illusory understanding (Lemke, 1990), triadic exchanges can mediate meaning (Nassaji & Wells, 2000). This paper focuses on one case from a study of discursive practices of seven middle-grades teachers identified for their expertise in mathematics instruction (Truxaw, 2004). The central result of the study was the development of a model to explain how teachers use discourse to mediate mathematical meaning in whole-group instruction. Drawing on the model for analysis, thick descriptions of one teacher's skillful orchestration of triadic exchanges that enhance student understanding of mathematics are presented.

**KEYWORDS**: communication; discourse analysis; language and mathematics; sociocultural theory

*The Role of Mathematics Discourse in Producing Leaders of Discourse*, pages 129–151

## INTRODUCTION

Recent reform efforts have identified communication as essential to the teaching and learning of mathematics (NCTM, 2000). However, the mere presence of talk does not ensure that thinking and understanding follow. Research has demonstrated that the *quality* and *type* of discourse greatly affect its potential for promoting conceptual understanding (Chapin, O'Connor, & Anderson, 2003; Kazemi & Stipek, 2001; Lampert & Blunk, 1998; Whitenack & Yackel, 2002) and that the teacher's role is vital in the orchestration of discourse (Barnes, 1992; Confrey, 1995; Jaworski, 1997; Nathan & Knuth, 2003). This paper focuses on one case from a larger study of discursive practices of seven-middle grades mathematics teachers identified for their expertise in mathematics instruction (Truxaw, 2004). The question addressed is: How does an expert middle-grades mathematics teacher orchestrate whole-group discourse to mediate mathematical meaning?

## BACKGROUND

Sociocultural theory, with its contention that higher mental functions derive from social interaction, provides a meaningful framework for analysis and discussion of discourse as a mediating tool in the learning-teaching process. Verbal exchanges between more mature and less mature participants may develop back and forth processes from thought to word and from word to thought that allow learners to move beyond what would be easy for them to grasp on their own (Vygotsky, 1934/2002, 1978). Instructional discourse should stretch toward the student's *potential* understanding (rather than the existing understanding), attending to his/her zone of proximal development (ZPD) (Vygotsky, 1978; Wertsch, 1991). These processes may promote semiotic activity—that is, mental activity related to signs and symbols (including verbal) that are used to create and communicate meaning (Lemke, 1990; van Oers, 2000). Semiotic mediation has the potential to transfer "psychological operation to higher and qualitatively new form" (Vygotsky, 1978, p. 40).

Related to semiotic mediation, Sfard (2000) describes discursive mechanisms that bring about the emergence of mathematical objects—that is, metaphorical objects that are "stable, permanent, self-sustained, and located beyond the discourse itself" (p. 322). In particular, she describes discursive foci that include *pronounced* focus that is public, *intended* focus that is mainly private, and *attended* focus that mediates between the two. In mathematics classrooms, strategic conversations may provide opportunities for semiotic mediation that may promote the development of "mathematical objects" and mathematical meaning-making.

When examining language as a mediator of meaning, it is useful to consider the two main functions of communication—that is, "to produce a maximally accurate transmission of a message" and "to create a new message in the course of the transmission" (Lotman, 2000, p. 68), characterized as *univocal* and *dialogic* discourse, respectively (Wertsch, 1998). Univocal discourse could be imagined with a conduit metaphor, with knowledge being sent in one direction. In contrast, dialogic discourse involves dialogue between at least two voices (Bakhtin, 1981) where the communication is used as a thinking device and new meaning is generated. Effective communication seems to require both univocal and dialogic aspects, a mixture of common understanding and different perspectives (Wertsch, 1998).

In addition to functions of discourse, researchers have identified basic structures found particularly within classrooms (e.g., Berry, 1981; Halliday, 1978; Mehan, 1985; Sinclair & Coulthard, 1975; Wells, 1999). For example, classroom discourse has been parsed according to the following categories: a move, an exchange, a sequence, and an episode (Wells, 1999). The move, exemplified by a question or an answer from one speaker, is identified as the smallest building block. An exchange, made up of two or more moves, occurs between speakers. Typically, the teacher initiates an exchange, with the student responding, and the teacher following-up with either an evaluation or some sort of feedback to the student's response. Exchanges are categorized as either nuclear (i.e., can stand alone) or bound (i.e., dependent upon or embedded within previous exchanges). The sequence is the unit that contains a single nuclear exchange and any exchanges that are bound to it. Finally, an episode comprises all the sequences that are necessary to complete an activity.

The most common pattern of classroom discourse follows the three-part exchange of teacher initiation, student response, and teacher evaluation or follow-up (i.e., IRE or IRF) (Cazden, 2001; Coulthard & Brazil, 1981; Mehan, 1985). The triadic structure has been criticized as encouraging "illusory participation"—that is, participation that is "high on quantity, low on quality"—because "it gives the teacher almost total control of classroom dialogue and social interaction" (Lemke, 1990, p. 168). However, Nassaji and Wells (2000) found that, even within inquiry-style instruction, triadic dialogue was the dominant structure and, therefore, important to take into account when examining classroom discourse. Further, it was noted that within triadic exchanges, the teacher's initiation and follow-up moves influence the function of the discourse. For example, when moves are used to evaluate a student's response, the intention of the discourse is likely to tend toward transmitting meaning (i.e., tending toward univocal). In contrast, questions that invite students to contribute ideas that might change or modify a discussion are more likely to be associated with dialogic discourse (Wells, 1999; Wertsch, 1998).

Along with sociocultural theory and classroom discourse research, socio-linguistics provides another lens for viewing and making sense of the flow, forms, functions, and intentions of classroom talk. For example, line-by-line coding strategies used in this study were adapted from Wells (1999) to enhance the analysis of classroom discourse transcripts. Additionally, forms of talk and verbal assessment within whole-group discussion were identified (Truxaw, 2004; Truxaw & DeFranco, 2004; Truxaw & DeFranco, 2008), including: *monologic talk* (i.e., involves one speaker—usually the teacher—with no expectation of verbal response), *leading talk* (i.e., occurs when the teacher controls the verbal exchanges, leading students toward the teacher's point of view), *exploratory talk* (i.e., speaking without answers fully intact, analogous to preliminary drafts in writing) (Cazden, 2001), *accountable talk* (i.e., talk that requires accountability to accurate and appropriate knowledge, to rigorous standards of reasoning, and to the learning community) (Michaels, O'Connor, Hall, & Resnick, 2002), *inert assessment* (IA) (i.e., assessment that does not incorporate students' understanding into subsequent moves, but rather, guides instruction by keeping the flow and function relatively constant), and *generative assessment* (GA) (i.e., assessment that mediates discourse to promote students' active monitoring and regulation of thinking (metacognition, Flavell, 1976, 1979) about the mathematics being taught).

The authors' previously reported research (Truxaw, 2004; Truxaw & DeFranco, 2008) demonstrated that graphic maps of discourse (called sequence maps) could be developed to represent the flow of the talk and verbal assessment moves and the overall function of the discourse (i.e., tending toward univocal or dialogic). Additionally, the sequence maps, when analyzed in conjunction with transcripts and other evidentiary data (e.g., interview transcripts and field notes), could be used to develop associated models of teaching. One of these models described *inductive teaching*—that is, teaching that moves from specific cases, through recursive cycles, toward more general hypotheses and rules (Truxaw, 2004).

## METHODS AND PROCEDURES

This paper focuses on one episode within one case from within a larger study. The participants in the larger study were a purposive sample of seven middle-grades mathematics teachers (grades 4–8 ) who were identified as having characteristics indicative of expertise (Darling-Hammond, 2000), including representatives from three specific groups: teachers who had achieved *National Board for Professional Teaching Standards* (NBPTS) certification in Early Adolescent Mathematics, recipients of the *Presidential Award for Excellence in Mathematics and Science Teaching* (PAEMST), and teachers recommended by university faculty. Discursive practices of one of these participants, Mr. Larson (all names are pseudonyms), an eighth-grade mathematics teacher, are highlighted in this paper. Mr. Larson's background included several indicators of expertise: 35 years of public school teaching experience, certification in secondary mathematics (grades 7–12), NBPTS certification, advanced certification in educational technology, bachelor's degree in economics, master's degree in mathematics education, and university-level mathematics education teaching experience.

The data were collected via semi-structured interviews, classroom observations, field notes, audiotapes, and videotapes. Grounded theory methodology (Strauss & Corbin, 1990), multiple-case study design (Stake, 1995; Yin, 1994), and sociolinguistic tools (Wells, 1999) were applied within a sociocultural framework (Vygotsky, 2002; Wertsch, 1991, 1998) to analyze the discourse. The transcripts from the classroom observations were coded on several levels—for example, line-by-line coding of moves was accomplished using schemes adapted from Wells (1999) (e.g., see Table 1) and sequences were coded using strategies developed during a pilot study. Next, individual *sequence maps* (i.e., diagrams representing the flow of forms of talk and verbal assessment within a sequence—see Figure 1) were created by applying the coded data to a preliminary graphic model of classroom discourse. Maps and coded transcripts were inspected, compared, adapted, and synthesized to develop an overall model of the flow of classroom discourse.

The central result of the larger study was the development of a dynamic model to explain how teachers use discourse to mediate mathematical meaning in whole-group instruction (see Figure 2) (Truxaw, 2004; Truxaw & DeFranco, 2004, Truxaw & DeFranco, 2008). The model provided a template for mapping the flow of forms of talk (i.e., monologic, leading, exploratory, and accountable) and forms of verbal assessment (i.e., IA and GA), along with tendencies toward univocal (transmitting meaning) or dialogic (generating meaning) function. Through its focus on the interactions of talk and verbal assessment and their relationships toward univocal and dialogic tendencies, the model supplied indicators to whether meaning was being conveyed or being constructed. Building a model of the flow of discourse provided a theoretical foundation and analytic tools to investigate how an expert mathematics teacher orchestrates whole-group discourse to mediate mathematical meaning.

To unpack how discourse could be orchestrated to mediate mathematical meaning, fine-grained analysis of the individual cases was undertaken. Certain sequences and instructional episodes were identified as potentially informative—with particular focus on sequences that included the following:

**TABLE 1. Example of Line-by-Line Coding Adapted from Wells (1999)**

| Map # | Line # | Seq # | Who | Text | Exch | Move (I-R-F) | K1 K2 | Prospectiveness | Function |
|---|---|---|---|---|---|---|---|---|---|
| 3 | 49 | 4 | T | What are the factors of six? B9? | Nuclear | I | K1 | Demand | Req inf |
| 4 | 50 | 4 | B9 | 2, 3, 1, and 6 | Dep | R | K2 | Give | Inform |
| 5 | 51 | 4 | T | Okay. | Dep | F | K1 | Acknow | Acknow |

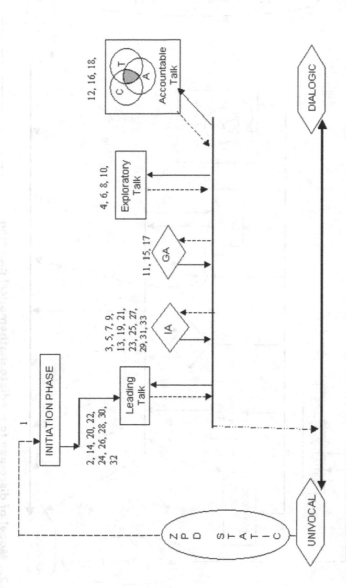

**FIGURE 1. Example of a sequence map representing the flow of particular forms of talk and verbal assessment. The numbers represent individual verbal moves in sequential order.**

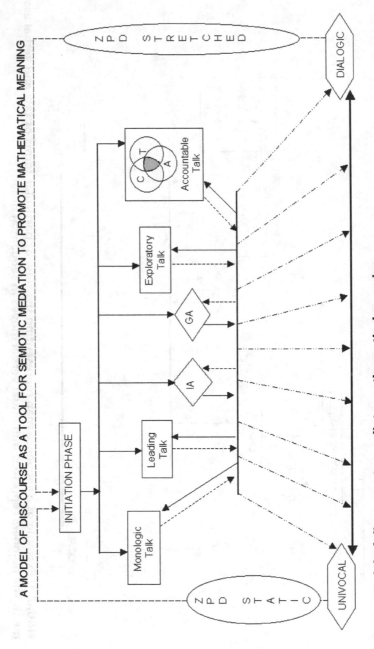

**FIGURE 2. Model of discourse to mediate mathematical meaning.**

GA = Generative Assessment; IA = Inert Assessment; C = Accountable to Community; A = Accountable to Accuracy; T = Accountable to Thinking. The model begins with an initiation phase, from which the discourse moves to a category of talk or verbal assessment. The type of verbal assessment is critical to the movement of discourse. IA provides little or no semiotic mediation and, therefore, tends to maintain existing discursive functions. GA promotes semiotic mediation and discourse is more likely to progress toward dialogic function.

- evidence of discourse that tended toward univocal function;
- evidence of discourse that tended toward dialogic function; and
- evidence of discourse that tended toward dialogic function but then shifted back toward univocal function.

Multi-level analysis of sequence maps, lesson transcripts, interview transcripts, and field notes of these selected episodes and sequences was completed. These analyses resulted in the development of models of teaching that demonstrated greater or lesser tendencies toward meaning-making. This paper reports on findings related to the analysis of one teaching/ learning episode from Mr. Larson's eighth grade mathematics class that was representative of an *inductive teaching model* (see Figure 3) and also demonstrated tendencies toward promoting meaning-making.

## RESULTS AND DISCUSSION

*Overview of an Inductive Teaching/ Learning Episode*

The inductive teaching model was built from a learning episode that consisted of four sequences in one of Mr. Larson's eighth grade mathematics classes. A problem was introduced in sequence one (i.e., "What is the sum of the reciprocals of the prime or composite factors of 28?"), establishing a frame of reference (see Figure 3-A). In sequence two, common understanding of key terms was developed (e.g., prime and composite) (see Figure 3-B-1), while in sequence three, the problem was investigated in small groups and a solution was presented by a student (see Figure 3-B-2 and 3). By consensus, the class agreed that the sum of the reciprocals of the prime and composite factors of 28 equaled 1, which provided a new basis for meaning-making (See Figure 3-C). Within the first three sequences, all four forms of talk and both IA and GA were used, but the overall function of the discourse was univocal in nature.

Although one might imagine that the learning episode would be complete with the presentation and acceptance of a solution, instead, the first three sequences served as a springboard for sequence four. The fourth sequence built from the first three, using the problem as a frame of reference for developing, testing and revising hypotheses; exploring connections between the problem's solution and other concepts (e.g., abundant numbers, deficient numbers, and perfect numbers); constructing revised frames of reference; and demonstrating students' understanding related to the original problem and the revised hypotheses. The cyclic nature of the discourse (i.e., recursively establishing common understanding, exploring, conjecturing, testing, and revising hypotheses) was used to progressively

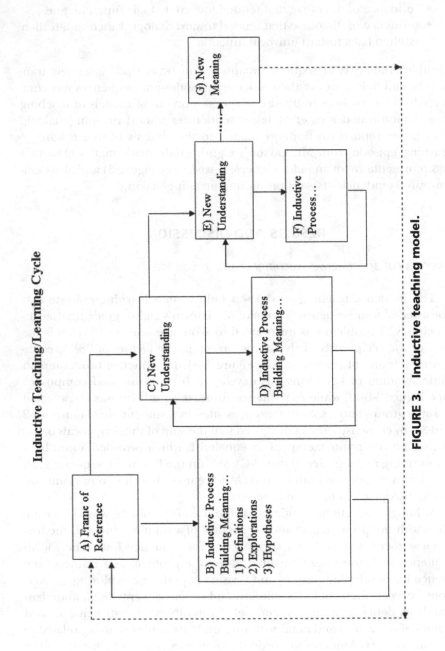

**FIGURE 3. Inductive teaching model.**

build new meaning (see Figure 3D-3G). The discourse in the fourth sequence was particularly complex—including multiple instances of leading, exploratory, and accountable talk and both IA and GA. Also of note is that accountable talk and GA occurred more frequently in sequence four than they had in the previous three sequences. The method of instruction was predominantly inductive—that is, moving from specific cases toward more general hypotheses and rules. The discourse in the learning episode (i.e., sequences 1– 4) moved from relatively univocal (while building common understanding) to relatively dialogic (as the common themes were used to build new meaning).

## Taking a Closer Look at the Episode

The fine-grained analysis of this teaching/learning episode (that resulted in the development of the inductive teaching model) included re-examination of the coded classroom transcripts, the sequence maps, pre- and post-observation interview transcripts, and field notes. These provided evidence for considering how the content, the flow, and the intent might influence the functions and outcomes of the discourse.

### Sequence One Revisited

In the first sequence, classic triadic exchange structure (IRE/IRF), as represented by monologic talk, leading talk, and IA, was used to introduce the problem (i.e., "What is the sum of the reciprocals of the prime or composite factors of 28?") that would serve as a frame of reference for the following three sequences. Mr. Larson derived the problem from one that his students had been given in a mathematics problem-solving competition the previous week—that is, "Find the sum of the reciprocals of all the factors of 28" (i.e., $1/1 + 1/2 + 1/4 + 1/7 + 1/14 + 1/28 = 56/28 = 2$). In an excerpt from Mr. Larson's post-observation interview, he described how his own curiosity about the problems from the math competition led him to the "guided discovery" (see glossary in Appendix) teaching/learning episode reported:

> I particularly liked the first problem and I was surprised at the result, (and) said, "Wow! If you take one off of here, you get one. Is that always going to be true—the reciprocals of the factors?" And I happened to choose six as the next number I tried and it worked again ... I said, (*inaudible*) ... we've got some funny ... it's a *perfect number...* then I said, oh, yeah, 28's a perfect number too. What if it's not a perfect number? I tried another number and it didn't work. And I realized, you know, it only works for perfect numbers. And I thought about how it made it sense. And wouldn't it be neat if kids could come to that ... themselves? So...I tried to massage the problem to make it

to be that ... and thought it might lead to a nice discussion about it ... (Post-observation interview, October, 2003)

Mr. Larson added, "It's so important that rather than just go into an answer key, and you can't always do it, but whenever it's possible, that you go through the experience that your kids have." Mr. Larson's process of working through the mathematics and *making his own discoveries* helped him to facilitate guided discovery with his students, as will be shown in the explanations of sequences two, three, and four.

### Sequence Two Revisited

Like the first sequence, the discourse in the second sequence was characterized by triadic exchanges between Mr. Larson and the students. However, this sequence was more complex than the first, including examples of leading, exploratory, and accountable talk and both IA and GA. The overall purpose of the sequence seemed to be the development of a common understanding of the vocabulary necessary for the problem—that is, prime and composite numbers. To establish this common language, Mr. Larson did not simply define the words; rather, he allowed students to express their own understanding, using both IA and GA to promote movement toward accurate and appropriate knowledge. For example, when a student offered an incomplete definition for prime numbers, "Numbers that can only be divided by one and itself," he asked for agreement from the class. When students agreed, he continued, facilitating examples and counterexamples, until a student provided the missing piece: "It has exactly two factors." Instead of simply accepting this, he used GA to probe, "How is that different?" The student provided an example, "Well, because one can divide itself, but it only has one factor." Overall, discourse tended to be univocal in that its purpose was to convey accurate meaning about the vocabulary; however, there was some tendency toward dialogic function since students' voices were key parts of the meanings that were represented and, additionally, justifications and explanations were included in the dialogue.

### Sequence Three Revisited

The third sequence included whole-group discourse (that was fully coded) and small group interactions (that were not coded line-by-line, but were described within field notes). Most of the verbal discourse during this sequence diverged from triadic structure—with a more open-ended, exploratory stance. During the small group interactions, Mr. Larson circulated throughout the room asking questions like, "How are you doing?" Although this type of question fell into the category of IA, Mr. Larson indicated (post-observation interview) that its purpose was open-ended to ensure that the students continued to explore the mathematical relationships inherent in the problem, as well as to provide a foundation for other assessments that would eventually lead toward the development of an understanding of the mathematics. During the small-group work, Mr. Larson listened and observed, identifying particular students who might need assistance and also students on whom he might call during large-group discussion to provoke meaningful discourse (post-observation interview). When the whole group reconvened, Mr. Larson selected a student to come to the board to demonstrate his strategies and share his solution. By consensus, the class agreed that the sum of the reciprocals of the prime and composite factors of 28 equaled 1. These first three sequences provided common understanding that served as a springboard from which sequence four was built.

### Sequence Four Revisited

Mr. Larson's fourth sequence was the most complex from among the 120 sequences that were mapped in the larger study. It included 6 leading talk moves, 10 exploratory talk moves, 28 accountable talk moves, 26 IA moves, and 20 GA moves. The discourse in this sequence moved back and forth between whole-group discourse (with predominantly triadic structure) and small group talk. During this sequence students were allowed to serve in the role of the *primary knower*, i.e., the person who "knows" the information and imparts it (Berry, 1981), the teacher modeled metacognitive reasoning (i.e., he modeled thinking out loud about his own reasoning processes) (Flavell, 1976, 1979), and new meaning was generated. Mr. Larson shifted his role throughout in the discourse—sometimes "stepping in" to serve as a participant in the discussion and sometimes "stepping out" to serve as a commentator who facilitated and clarified rules, procedures, concepts, arguments, and classroom norms (Rittenhouse, 1998). In sequence four, Mr. Larson orchestrated the introduction of a hypothesis related to the problem that was introduced in sequence one.

> Mr. Larson: That's sort of surprising that it would actually be one. I wonder if that's always true. I'm going to try another. I'm going to try six. Somebody said something about six. What are the factors of six? Bruce?
>
> Bruce: 2, 3, 1, and 6
>
> Mr. Larson: Okay. (Writes factors on board.) So, again, what we said was, we're going to only use the prime and composite factors, right? So we'll throw out this one. (Crosses out the 1.) So we have 1/2 plus 1/3 plus 1/6, right? And I like what David did, (references student work from the previous sequence) which was, he made a same denominator. (Shows 3/6 + 2/6 + 1/6 on board.) And what do we get?
>
> Lindsey: 6/6

> Mr. Larson: Which is *one*! Whoa! So what should we call this? Should we
>   call this the *Hankins* Hypothesis or what? (using David's last name)
>   You want credit for it, David?
> David: Definitely.
> Mr. Larson: Definitely. Well, I wonder if this always works. It looks kind
>   of neat. We've seen two examples now where it works. I'm sort of
>   surprised…I don't know why it would work, but it seems to work.
> Mr. Larson: Would you guys check it out? Would you each take some
>   other number and check it to see if, in fact, it does work?

The students worked cooperatively in small groups at tables, testing out
numbers. During this time, Mr. Larson wrote the "Hankins Hypothesis"
on the board—that is: "Sum of reciprocals of prime and composite factors
of a number will always be one." Mr. Larson then circulated around the
room, listening and asking questions. For example, when a student said,
"It doesn't work for primes," Mr. Larson asked him to think about why that
might be so. After calling the class's attention back to the whole group,
Mr. Larson asked, "Okay, so what did you discover?" As the students shared
their "discoveries," Mr. Larson orchestrated the addition of two potential
exceptions to the Hankins Hypothesis that he named after the students
who uncovered them. They stated that "the hypothesis doesn't work for
primes" and "the hypothesis doesn't work for perfect cubes." Additional
numbers for which the Hankins Hypothesis didn't work were shared, until
a student said, "It didn't work for 36, which is an *abundant number*."

> Mr. Larson: Whoa! A what? (dramatically)
> David: An abundant number.
> Mr. Larson: An *abundant* number! What is an abundant number?
> David: When the factors of the number add up to more than the number
>   itself.
> Mr. Larson: Have you guys heard of that before? An *abundant* number?
> Students: (A couple of students indicate that they have)
> Mr. Larson: And do you know what it's called when a number, the sum of
>   the factors, add up to something less than the number?
> Kohei: A non-abundant number?
> Mr. Larson: Somebody … A non-abundant number … somebody … I
>   heard it out there.
> Lindsey: Deficient.
> Mr. Larson: Deficient. So, a *deficient* number and an *abundant* number…

The class then briefly discussed the meanings of abundant and deficient
numbers. Then a student made a connection between the new knowledge
of deficient numbers and the Hankins Hypothesis.

Mr. Larson: And, Kohei, you have a big smile on your face.

Kohei: It doesn't work for deficient numbers.

Mr. Larson: It doesn't work for deficient numbers. Hmmm. Daniel.

Daniel: Uh, you said, "abundandant." (Refers to spelling mistake on board.)

Mr. Larson: Abundantant—I'm just making up a new word. So, huh? It doesn't work for abundant... Is there anything that we call it when the sum of the factors of a number itself *equals* the number?

David: Perfect numbers?

Mr. Larson: It's called a *perfect* number. It's called a *perfect* number.

Mr. Larson asked students how many had heard of perfect numbers, and then asked if anyone knew any perfect numbers. Students made connections between the original problem (from sequence one) and perfect numbers.

Mr. Larson: ... Perfect number. Well, anybody know any perfect numbers? Daniel?

Daniel: Six.

(Recall that the number 6 worked for the "Hankins Hypothesis.")

Mr. Larson: Six is a perfect number. Huh! ...and, Arthur?

Arthur: 28.

(Recall that 28 was the number introduced in the original problem.)

Mr. Larson: 28 is a perfect number. Mmmmmm.

Mr. Larson: David, would you like to modify the Hankins Hypothesis?

David: Uh, they have to be perfect numbers, not just any number.

Mr. Larson: Let's see here...So the sum of the factors of prime and composite (Reads from board as he adjusts the Hankins Hypothesis)... sum of the reciprocals of prime and composite factors of a perfect number will be one.

The sequence concluded with Mr. Larson challenging the students to find the next perfect number and to see if it fit the newly-revised Hankins Hypothesis.

### Summary of the Inductive Teaching/Learning Episode

Although classic triadic exchange structure and univocal discourse were clearly evident within the teaching/learning episode described (especially in sequences 1–3), there were also indicators that mathematical meaning-making took place (especially in sequence 4). Common language related to the initial problem was developed in the first three sequences. This mutually-built understanding served as a foundation for setting up exploration of richer concepts. Within sequence four, conjectures led to a preliminary hypothesis that Mr. Larson named after the student who demonstrated the initial problem's solution—that is, the sum of reciprocals of prime and composite factors of a number will always be one. Mr. Larson used discourse to model metacognitive processes (Flavell, 1976, 1979) ("… That's sort of surprising that it would actually be one. I wonder if that's always true. I'm going to try another …") and then proceeded to have students explore possibilities ("… We've seen two examples now where it works. I'm sort of surprised … I don't know why it would work, but it seems to work… Would you guys check it out? Would you each take some other number and check it to see if, in fact, it does work?"), including exceptions to the hypothesis ("… doesn't work for prime numbers …"). Mr. Larson's knowledge of mathematical content and pedagogy (Shulman, 2000) were instrumental in the orchestration of the discourse. He encouraged the students to explore and to conjecture, but also supplied meaningful verbal assessments that provoked them to generalize, to question, to justify, to reformulate, and to develop new meaning, thus including tendencies toward dialogic discourse.

It is important to note that the whole-group discourse that was identified as promoting meaning-making did not stand on its own—it was integrally connected to earlier instances of both whole-group and small-group discussions. For example, recursive cycles of whole-group and small-group discussion were used to establish a rich problem as a frame of reference, develop common language, explore the problem, and develop and test hypotheses. The whole-group meaning-making discourse explored connections between the problem's solution and other problems, built revised frames of reference, and demonstrated students' understanding (related to the original problem and revised hypotheses). Although the whole-group discourse included classic triadic discourse structures, it also included GA and accountable talk. Furthermore, opportunities for rich verbal interactions within whole-group discussion were built from preceding sequences that established a frame of reference, developed common language, and provided opportunities for exploration and conjecture. This case provides examples of how whole-group discourse can be used to mediate student understanding of mathematics.

## FINAL REMARKS

"Mathematics, when it is finished, complete, all done, then it consists of proofs. But, when it is discovered, it always starts with a guess…" (Pólya, 1966). In the Mathematics Association of America's video classic, "Let Us Teach Guessing," George Pólya, noted mathematician, mathematics educator, and problem-solving expert, can be seen teaching "guessing" to a group of university students. He begins the lesson with a rich problem that is unfamiliar to the students. As the lesson progresses, Pólya guides the students through a cycle of guesses, investigations, hypotheses, further investigations, and conjectures. Although no formal proofs are presented, evidence builds toward mathematical sense-making. Pólya makes use of triadic exchanges to facilitate the lesson, but does so with an art that seems to derive from deep understanding of both content and pedagogy (Truxaw & DeFranco, 2007). Mr. Larson, a teacher with 35 years of experience in both mathematics content and pedagogy, seemed to make similar use of triadic structures as he worked through his own inductive teaching/learning cycle with his students. In the episode described, triadic structures seemed related to meaning-making when the verbal exchanges were connected to rich mathematical problems and to building (rather than simply conveying) students' understanding.

In revisiting previously-stated concerns that triadic exchange structure may promote "illusory participation" (Lemke, 1990), we are reminded that consistent findings of classroom studies show that most US teachers do tend to *state* information rather than develop ideas with their students (NCES, 1999, 2001). However, current research provides evidence that simply engaging students more actively in classroom discourse is not a panacea for improving mathematical achievement. For instance, Nathan and Knuth (2003) investigated a sixth-grade teacher's attempts to become more reform-oriented by working to decrease her authority-role in discussions while increasing her students' participation. Although the teacher was able to accomplish her discursive goals, "When the teacher elected to move away from her analytic role, the team (of researchers) observed that there was nothing added to the classroom culture to fill this gap in the discourse when there were major oversights, or when conflicting views among students arose" (p. 200). One striking example was when the students opted to "vote" on a mathematical concept. Lobato, Clarke, and Ellis (2005) suggest that, although many reform-oriented teachers have downplayed "actions centered on introducing new mathematical ideas" (p. 104), it may be appropriate to reconsider "telling" as a "system of actions," as long as the teacher focuses attention on the "development of the students' mathematics rather than on the communication of the teacher's mathematics" (p. 109). Indeed, perhaps a key piece is the professional judgment of an experienced teacher to know when to shift roles—that is, when to "step in" as a participant and when to "step out" to become a commentator of rules, norms, and concepts (Rittenhouse, 1998).

Although triadic-exchange structure is often used to control instruction rather than to promote meaning-making, the episode from Mr. Larson's mathematics class described here provides evidence that these structures can also be part of an inductive model of teaching. In his teaching, Mr. Larson did include some judicious "telling" (Lobato et al., 2005), but also used his pedagogical content knowledge to judge when to "step in" and when to "step out" (Rittenhouse, 1998) in order to focus attention on the students' mathematical meaning-making. And, like George Pólya (1966), Mr. Larson invited students to guess, to hypothesize, to justify, and to make sense of mathematics. In summary, it is important to continue to explore all avenues of instruction that may invite students to make sense of mathematics. Even with its limitations, triadic exchange structure may still be used effectively by teachers whose learning/teaching goals are based on building students' understanding, rather than on simply conveying the teacher's ideas.

## APPENDIX—GLOSSARY OF TERMS
## (AS USED IN THIS PAPER)

Accountable Talk: Classroom talk where participants are accountable to accurate and appropriate knowledge, to rigorous standards of reasoning, and to the learning community (Institute for Learning, 2001; Resnick, 1999).

Assessment (verbal): For the purposes of this research, includes verbal moves (usually by the teacher) that help the teacher guide instruction and/or enhance learning.

Dialogic Discourse: Dialogue used to generate meaning (Knuth & Peressini, 2001; Lotman, 1988).

Discourse: "Purposeful talk on a mathematics subject in which there are genuine contributions and interaction" (Pirie, 1998; Pirie & Schwarzenberger, 1988, p. 460).

Episode: Largest unit in classroom discourse: all the talk to carry out a single activity (Wells, 1999).

Exchange: Two or more moves occurring between speakers—typically structured as initiation, response, (and often) follow-up (IRF) (Wells, 1999).

Exploratory Talk: "Speaking without answers fully intact (Cazden, 1988/2001, p. 170). Exploratory talk is analogous to first drafts in writing, that is, first steps toward fluent and elaborated talk (Barnes, 1992).

Generative Assessment [GA]: Assessment that mediates discourse to promote students' active monitoring and regulation of thinking (i.e., metacognition) about the mathematics being taught (e.g., involvement that provokes elaboration and reflection, like, "What do you think?" or "Why do you think that?"). GA often changes the flow and function of the discourse (Truxaw, 2004).

Guided Discovery: Teaching that guides students to develop understanding of concepts through to inquiry and problem solving (Bruner, 1973; Pólya, 1962/1981).

Inductive Teaching: Teaching that moves from speci c cases, through recursive cycles, toward more general hypotheses and rules (Truxaw, 2004). Pólya (1962/1981) advocated that mathematics instruction be facilitated through phases of exploration, formalization, and assimilation.

Inert Assessment [IA]: Assessment that is likely to keep the flow and function of the discourse relatively constant (i.e., status quo), tending toward univocal (e.g., comments like, "Nice job," or, "That is not correct"). *Inert assessment* is based on Alfred North Whitehead's description of "inert ideas," that is, ones "that are merely received into the mind without being utilized, or tested, or thrown into fresh combinations" (Whitehead, 1964, p. 13).

IRF: Initiation, response, follow-up (Coulthard & Montgomery, 1981).

Leading Talk: Classroom talk in which the teacher controls the verbal exchanges, leading students toward the teacher's understanding. Although verbal exchanges occur, leading talk serves essentially the same purpose as monologic talk. Students' responses that have been *led toward* the teacher's intent are coded as leading talk (Truxaw, 2004).

Metacognition: "'Metacognition' refers to one's knowledge concerning one's own cognitive processes and products or anything related to them … Metacognition refers, among other things, to the active monitoring and consequent regulation and orchestration of these processes in relation to the cognitive objects on which they bear, usually in the service of some concrete goal or objective" (Flavell, 1976, p. 232).

Metacognitive Talk: Classroom talk in which participants actively monitor, regulate, and orchestrate their own thinking and learning processes. (The teacher provides an important role as he/she models/mediates/scaffolds this talk.)

Monologic Talk: Classroom talk in which one person (usually the teacher) is the only speaker. No verbal response is expected.

Move: Smallest building block in classroom discourse (Wells, 1999).

Semiotic Mediation: Instead of mental activity being entirely driven by environmental stimuli (e.g., stimulus: response), it is mediated by auxiliary stimuli (e.g., words). "Because this auxiliary stimulus possesses the specific function of reverse action, it transfers the psychological operation to higher and qualitatively new forms and permits humans, by their aid of extrinsic stimuli, *to control their behavior from the outside*" (Vygotsky, 1978, p. 40).

Semiotics: Semiotics is the study of all systems of signs and symbols and how they are used to communicate meanings (Lemke, 1990).

Sequence: Unit that includes a single nuclear exchange with any exchanges that are bound to it (Wells, 1999).

Sequence Map: A graphic map showing all forms of talk and assessment contained within a single sequence and the flow between them (Truxaw, 2004).

Triadic Structure: The most common structure of classroom discourse is *triadic dialogue* that includes IRE/IRF (initiation, response, evaluation/follow-up) (Cazden, 1988/ 2001; Knuth & Peressini, 2001).

Univocal Discourse: One-way transmission of meaning (Knuth & Peressini, 2001; Lotman, 1988).

Zone of Proximal Development [ZPD]: "The distance between the actual developmental level as determined by independent problem solving and the level of potential development as determined through problem solving under adult guidance or in collaboration with more capable peers" (Vygotsky, 1978, p. 86).

## REFERENCES

Bakhtin, M. M. (1981). *The dialogic imagination: Four essays by M. M. Bakhtin.* (M. Holquist, Ed.; C. Emerson & M. Holquist, Trans.). Austin: University of Texas Press. (Original work published in the 1930s)

Barnes, D. R. (1992). *From communication to curriculum* (2nd ed.). Portsmouth, NH: Boynton/Cook Publishers.

Berry, M. (1981). Systemic linguistics and discourse analysis: A multi-layered approach to exchange structure. In M. Montgomery (Ed.), *Studies in discourse analysis* (pp. 120-145). London; Boston: Routledge & Kegan Paul.

Bruner, J. S. (1973). *Beyond the information given: Studies in the psychology of knowing.* New York: Norton.

Cazden, C. B. (2001). *Classroom discourse: The language of teaching and learning* (2nd ed.). Portsmouth: Heinemann. (Original work published 1988)

Chapin, S. H., O'Connor, C., & Anderson, N. C. (2003). *Classroom discussions using math talk to help students learn, grades 1–6.* Sausalito CA: Math Solutions Publications.

Confrey, J. (1995). How compatible are radical constructivism, sociocultural approaches, and social constructivism? In J. Gale (Ed.), *Constructivism in education* (pp. 185–225). Hillsdale, NJ: Lawrence Ehrlbaum.

Coulthard, M., & Brazil, D. (1981). Exchange structure. In M. Montgomery (Ed.), *Studies in discourse analysis* (pp. 82–106). London; Boston: Routledge & Kegan Paul.

Coulthard, M., Montgomery, M., & Brazil, D. (1981). Developing a description of spoken discourse. In M. Coulthard & M. Montgomery (Eds.), *Studies in discourse analysis* (pp. 1–50). London ; Boston: Routledge & Kegan Paul.

Darling-Hammond, L. (2000). Teacher quality and student achievement: A review of state policy evidence. *Education Policy Analysis Archives, 8*(1).

Flavell, J. H. (1976). Metacognitive aspects of problem solving. In L. B. Resnick (Ed.), *The nature of intelligence* (pp. 231–251). Hillsdale, NJ: Lawrence Erlbaum Associates.

Flavell, J. H. (1979). Metacognitive and cognitive monitoring: A new area of cognition—Developmental inquiry. *American Psychologist, 34,* 906–911.

Halliday, M. A. K. (1978). *Language as social semiotic: The social interpretation of language and meaning.* Baltimore: University Park Press.

Institute for Learning. (2001). *Accountable talk: Classroom conversation that works.* University of Pittsburgh. Retrieved May 7, 2003, from the World Wide Web: http://www.instituteforlearning.org/at.html

Jaworski, B. (1997). Tensions in teachers' conceptualisations of mathematics and of teaching. Paper presented at the Paper presented at the Annual Meeting of the American Educational Research Association (Chicago, IL, March, 1997).

Kazemi, E., & Stipek, D. (2001). Promoting conceptual thinking in four upper-elementary mathematics classrooms. *Elementary School Journal, 102*(1), 59–80.

Knuth, E., & Peressini, D. (2001). Unpacking the nature of discourse in mathematics classrooms. *Mathematics Teaching in the Middle, 6,* 320–325.

Lampert, M., & Blunk, M. L. (1998). *Talking mathematics in school: Studies of teaching and learning.* Cambridge UK; New York: Cambridge University Press.

Lemke, J. L. (1990). *Talking science: Language, learning, and values.* Norwood, N.J.: Ablex.

Lobato, J., Clarke, D., & Ellis, A. B. (2005). Initiating and eliciting in teaching: A reformation of telling. *Journal for Research in Mathematics Education, 36,* 101–136.

Lotman, Y. (1988). Text within a text. *Soviet Psychology, 24,* 32–51.

Lotman, Y. M. (2000). *Universe of the mind: A semiotic theory of culture* (A. Shukman, Trans.). London: I. B. Tauris & Co Ltd. (Original work published in 1990)

Mehan, H. (1985). *The structure of classroom discourse. Handbook of discourse analysis* (Vol. 3): Academic Press.

Michaels, S., O'Connor, M. C., Hall, M. W., & Resnick, L. (2002). *Accountable talk: classroom conversation that works* (Abridged version of E-book excerpted from Beta version 2.1 CD-Rom series). Pittsburgh: University of Pittsburgh.

Nassaji, H., & Wells, G. (2000). What's the use of "triadic dialogue"?: An investigation of teacher-students interaction. *Applied Linguistics, 21,* 376–406.

Nathan, M. J., & Knuth, E. J. (2003). A study of whole classroom mathematical discourse and teacher change. *Cognition and Instruction, 2,* 175–207.

NCES. (1999). *The TIMSS videotape classroom study: Methods and findings from an exploratory research project on eighth-grade mathematics instruction in Germany, Japan, and the United States* (No. NCES 99-074). Washington, D.C.: National Center for Education Statistics, U.S. Department of Education.

NCES. (2001). *Highlights from the Third International Mathematics and Science Study-Repeat (TIMSS-R)* (No. NCES-2001-027). Washington, D. C.: National Center for Education Statistics.

NCTM. (2000). *Principles and standards for school mathematics.* Reston, VA: National Council of Teachers of Mathematics.

Pirie, S. (1998). Crossing the gulf between thought and symbol: Language as (slippery) stepping-stones. In M. Bartolini-Bussi, A. Sierpinska, and H. Steinbrig (Eds.), *Language and communication in the mathematics classroom* (pp. 7–29). Reston VA: National Council of Teachers of Mathematics.

Pirie, S. E. B., & Schwarzenberger, R. L. E. (1988). Mathematical discussion and mathematical understanding. *Educational Studies in Mathematics, 19*, 459–470.

Pólya, G. (1962/1981). *Mathematical discovery: On understanding, learning, and teaching problem solving.* New York: John Wiley & Sons.

Pólya, G. (1966). Let us teach guessing: A demonstration with George Pólya [Motion picture] [Video recording]. In K. Simon (Producer), *MAA video classics.* Washington, D.C.: Mathematical Association of America.

Resnick, L. (1999). *Making America smarter.* Education Week on the Web. Retrieved May 8, 2003, from the World Wide Web: http://www.edweek.org/ew/vol-18/40resnick.h18 bv

Rittenhouse, P. S. (1998). The teacher's role in mathematical conversation: stepping in and stepping out. In M. L. Blunk (Ed.), *Talking mathematics in school: Studies of teaching and learning* (pp. 163–189). Cambridge UK; New York: Cambridge University Press.

Sfard., A. (2000). Steering (dis)course between metaphors and rigor: Using focal analysis to investigate an emergence of mathematical objects. *Journal for Research in Mathematics Education, 31*, 296–327.

Sinclair, J. M., & Coulthard, R. M. (1975). *Towards an analysis of discourse: The English used by teachers and pupils.* London: Oxford University Press.

Stake, R. E. (1995). *The art of case study research.* Thousand Oaks: Sage Publications.

Strauss, A. L., & Corbin, J. M. (1990). *Basics of qualitative research: Grounded theory procedures and techniques.* Newbury Park, Calif.: Sage Publications.

Truxaw, M. P. (2004). Mediating mathematical meaning through discourse: An investigation of discursive practices of middle grades mathematics teachers. (Doctoral dissertation, University of Connecticut, 2004). *Dissertation Abstracts International, 65*(08), 2888B.

Truxaw, M. P. & DeFranco, T. C. (2004, October). A model for examining the nature and role of discourse in middle grades mathematics classes. *Proceedings of the 26th Annual Meeting: North American Chapter of the International Group for the Psychology of Mathematics Education, 2*, 805–813.

Truxaw, M. P., & DeFranco, T. C. (2007). Mathematics in the making: Mapping verbal discourse in Pólya's "let us teach guessing" lesson. *Journal of Mathematical Behavior, 26*(2), 96–114.

Truxaw, M. P., & DeFranco, T. C. (2008). Mapping mathematics classroom discourse and its implications for models of teaching. *Journal for Research in Mathematics Education, 39*, 489–525.

van Oers, B. (2000). The appropriation of mathematical symbols: A psychosemiotic approach to mathematics learning. In P. Cobb, E. Yackel, & K. McLain (Eds.) *Symbolizing and communicating in mathematics classrooms: Perspectives on discourse tools, and instructional design* (pp. 133–176). Mahwah N J: Lawrence Erlbaum Associates.

Vygotsky, L. S. (1978). *Mind in society: The development of higher psychological processes* (M. Cole, V. John-Steiner, S. Scribner, & E. Souberman, Eds.). Cambridge, MA: Harvard University Press.

Vygotsky, L. S. (2002). *Thought and language* (13th ed.) (A. Kozulin, Ed. & Trans.). Cambridge, MA: MIT Press (Original work published 1934).

Wells, G. (1999). *Dialogic inquiry: Toward a sociocultural practice and theory of education.* New York: Cambridge University Press.

Wertsch, J. V. (1991). *Voices of the mind: A sociocultural approach to mediated action.* Cambridge, MA: Harvard University Press.

Wertsch, J. V. (1998). *Mind as action.* New York: Oxford University Press.

Whitehead, A. N. (1964). *The aims of education: And other essays.* New York: New American Library.

Whitenack, J., & Yackel, E. (2002). Making mathematical arguments in the primary grades: The importance of explaining and justifying ideas. *Teaching Children Mathematics, 8,* 524–527.

Yin, R. K. (1994). *Case study research: Design and methods* (2nd ed.). Thousand Oaks: Sage Publications.

# CHAPTER 8

# ELICITING HIGH-LEVEL STUDENT MATHEMATICAL DISCOURSE

## Relationships between the Intended and Enacted Curriculum

### Nicole Miller Rigelman

Striving for deepened student mathematical discourse, this study examines the mathematical discourse in the intended and the enacted curriculum in project teachers' classrooms. Using a continuum of types of student mathematical discourse, the researcher inferred the levels of student mathematical discourse as prompted by the written curriculum materials (the intended curriculum) and assessed the levels of discourse evident in videotapes of third and fourth grade lessons (the enacted curriculum). The teachers and researcher collaboratively examined and planned lessons with a focus on increasing students' ability to explain, question, challenge, conjecture, justify, and generalize mathematical ideas. The research team was interested in knowing more about why higher levels of discourse seemed to be less likely in early portions of a unit of study and they planned with a press for higher level thinking and reasoning. While the level of student mathematical discourse may vary across a unit of study and be dependent on students' prior experiences and/or familiarity with content, this study finds the norms of the classroom learning environment to be a highly critical factor in determining the level of student mathematical discourse. High-level student mathematical

*The Role of Mathematics Discourse in Producing Leaders of Discourse*, pages 153–172
**153**

discourse is not only possible but highly probable in classrooms where appropriate support and clear expectations are in place.

**KEYWORDS:** mathematical discourse, curriculum implementation, problem-solving, reasoning, sense-making, classroom environment

## INTRODUCTION

Teachers establish and nurture an environment conducive to learning mathematics through the decisions they make, the conversations they orchestrate, and the physical setting they create. Teachers' actions are what encourage students to think, question, solve problems, and discuss their ideas, strategies, and solutions. The teacher is responsible for creating an intellectual environment where serious mathematical thinking is the norm (NCTM, 2000, p.18).

The *Principles and Standards for School Mathematics* (NCTM, 2000) offer a vision for school mathematics where students work to make sense of mathematical ideas. Students do this by getting inside the subject, seeing how things work, and connecting existing knowledge and ideas to their new learning (Battista, 2007; Hiebert, Carpenter, Fennema, Fuson, Wearne, & Murray, 1997). Teachers support their students by creating a mathematically-productive learning environment, focused on understanding and sense-making for each student, mathematical discourse becomes a tool for "making public" students' strategies, reasoning, and solutions.

Numerous researchers have found that curriculum materials strongly influence teachers' actions and decision-making (Remillard, 1999; Rigelman, 2003; Drake & Sherin, 2006) suggesting that curriculum use also influences the student mathematical discourse within a lesson. This study examines the mathematical discourse in the intended and the enacted curriculum in project teachers' classrooms. Curriculum enactment is directly influenced by, among other things, teachers' mathematical content knowledge and specifically the mathematical knowledge needed for teaching (Hill, Schilling, & Ball, 2004).

A major goal of the Oregon Mathematics Leadership Institute (OMLI)[1] partnership is to develop teachers' mathematical knowledge through engaging with graduate-level mathematics in a research-based best practices learning environment i.e., the OMLI summer institutes. Teachers also participated in leadership courses extended through school-year site vis-

[1]The Oregon Mathematics Leadership Institute Partnership Project is funded by the National Science Foundation's Math Science Partnership Program (NSF-MSP award #0412553) and through the Oregon Department of Education's MSP program.

its, designed to support them in developing and sustaining collaborative professional learning communities with a focus on improving student understanding and achievement. As teachers learned about leadership and engaged personally in various practice-based professional development strategies and thought through the facilitation of such activities, they continued to deepen their understanding about best practices in teaching mathematics. In turn, as teachers reformed their practices to reflect best practices and worked with their colleagues to do the same, the hope was that they would positively impact student achievement. The OMLI project leaders hypothesized that by increasing the quantity and quality of student mathematical discourse, student achievement in mathematics could be significantly improved.

## MATHEMATICAL DISCOURSE AND COGNITIVE DEMAND OF TASK

The discourse of a classroom—the ways of representing, thinking, talking, agreeing, and disagreeing—is central to what and how student learn about mathematics (NCTM, 2007, p. 46).

Mathematics discussions are key to current visions of effective mathematics teaching and learning (Hiebert, et. al., 1997; NCTM, 2000, 2007; Sherin, 2002). As noted above, mathematical discussion influences *what* students learn—the knowledge they construct, and *how* students learn—the discourse practices they adopt. Sherin (2002) describes discourse structures teachers employ that influence the level of mathematical discourse and are mediated by the teachers' goals for the lesson.

**TABLE 1. Discourse Types**

| | |
|---|---|
| Type 1: | Answering, stating, or sharing: A student gives a short right or wrong answer to a direct question or makes a simple statement or shares work that does not involve an explanation of how or why. |
| Type 2: | Explaining: A student explains a mathematical idea or procedure by describing how or what he or she did but does not explain why. |
| Type 3: | Questioning or challenging: A student asks a question to clarify his or her understanding of a mathematical idea or procedure or makes a statement or asks a question in a way that challenges the validity of an idea or procedure. |
| Type 4: | Relating, predicting or conjecturing: A student makes a statement indicating that he or she has made a connection or sees a relationship to some prior knowledge or experience or makes a prediction or a conjecture based on an understanding of the mathematics behind the problem. |
| Type 5: | Justifying or generalizing: A student provides a justification for the validity of a mathematical idea or procedure or makes a statement that is evidence of a shift from a specific example to the general case. |

Note: From Weaver & Dick (2006).

Taking up the idea that the level of discourse influences what students learn, OMLI project staff defined five types of student discourse (see Table 1). These discourse types represent a continuum of the mathematical discourse present in mathematics classrooms where students are thinking about and talking about mathematics. The order of the discourse types, as presented in the table, represents increasing levels of cognitive demand. That is, giving a short right or wrong answer to a direct question represents the lowest level of cognitive demand and justifying mathematical ideas and procedures and making generalizations represent the highest levels.

Informing the design of the OMLI mathematics and leadership courses was the *Mathematics Tasks Framework* (see Figure 1) (Smith & Stein, 1998; Stein, Grover, & Henningsen, 1996; Stein, Smith, Henningsen, & Silver, 2000). This framework was used to analyze the enactment of hundreds of lessons. Findings indicate that:

1.  high-level cognitive demand tasks are difficult to implement well and often decline in demand during instruction; and
2.  students' learning was greatest in classrooms where tasks consistently encourage high-level thinking and reasoning (Stein, Smith, Henningsen, & Silver, 2000) and as noted previously, the reasoning is made public through the mathematical discourse that occurs.

With the belief that high-level cognitive demand tasks are key to deepening learning, project teachers learned through worthwhile mathematical tasks implemented with strategies that promoted high level thinking and

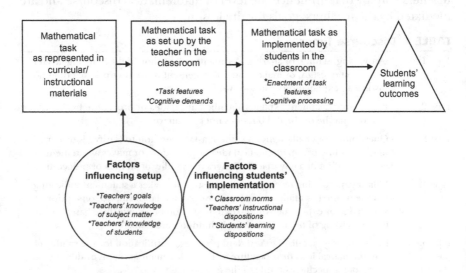

**FIGURE 1.   From Stein, Grover, & Henningsen (1996).**

discourse. Teachers learned about this pedagogy through analyzing these learning experiences and having explicit conversations about the instructional moves and ways that they support or hindered individual's learning. In turn, teachers would "try on" these best practices both in their own classroom and within the school-based professional learning communities.

## ELICITING HIGH-LEVEL STUDENT MATHEMATICAL DISCOURSE

Teachers should consistently expect students to explain their ideas, to justify their solutions, and to persevere when they encounter difficulties. Teachers must also help students learn to expect and ask for justifications and explanations from one another (NCTM, 2007, p. 40).

A key role for the teacher is establishing and maintaining an environment conducive of problem solving and inquiry. Much is communicated through a teacher's moves, such as availability of tools, press for justification, and respect for students' ideas. Each move establishes norms for classroom interaction that also influence students' efforts (Hiebert, et. al., 1997; Hufferd-Ackles, Fuson, & Sherin, 2004; NCTM, 2007, Sherin, 2002).

More recently, the work of Stein and her colleagues has been extended to a protocol for lesson planning (Stein, Engle, Hughes, & Smith, under consideration) that supports a teacher in thinking through the implementation of a high-level cognitive demand task and "controlling" the content and level of student thinking and discourse that emerge from a task. They identify five key planning practices:

1.  anticipating;
2.  monitoring;
3.  selecting;
4.  sequencing; and
5.  connecting;

suggesting that these planning practices minimize the level of improvisation the teacher must engage in while moving discussion in mathematically productive directions. Project teachers regularly engage in collaborative lesson planning using this tool to support their goals for continued press for justification and generalization of mathematical ideas.

## BACKGROUND

The early discourse data from OMLI project revealed that the discourse data is "flat" for the OMLI summer institute (i.e., there is about the same number of instances per hour of each of the various discourse levels). After working with teachers during site visits in year one, this researcher wondered if "flat" discourse data is possible in elementary classrooms. Teachers felt that it was not until the second or third day of a lesson or was not until the end of a unit that students would engage in higher levels of mathematical discourse. From this, questions emerged such as:

1. Does it take more than one class session for a teacher to get to the point in the lesson where higher levels of discourse occur among students[2]?
2. Is a transition to higher-level discourse delayed because the curriculum materials do not prompt for the higher-level discourse?
3. What is possible from students, in terms of high-level thinking and reasoning at early points in the unit?

It is these questions that this paper attempts to answer.

## METHODS AND CONTEXT

To examine the relationship between the intended and the enacted curriculum, the researcher analyzed the lessons in the curriculum relative to the levels of student mathematical discourse. This analysis included a review of the cognitive demand of the task(s) in the lesson (Henningsen & Stein, 1997; Smith & Stein, 1998; Stein, Smith, Henningsen, & Silver, 2000), the pedagogical structures of the lesson/curriculum (Hiebert, et. al., 1997) and/or ways in which discourse is supported, and how the lesson is situated in the larger mathematical storyline across the unit and/or the school year as this seems to influence student readiness for conjecturing and generalizing.

The four teachers were among the 180 teacher leaders participating in the OMLI project. They represented three schools that were part of the OMLI project and were located in two suburban districts. The districts each have *Investigations in Number, Data, and Space* (Mokros & Russell, 1995; Russell & Economopoulos, 2008) as their adopted curriculum at the third through fifth grade level. Selected OMLI teachers were invited to take part in this study, each choosing to participate as they saw that the collaborative work provided her with a professional development opportunity. The teachers and the researcher engaged in collaborative lesson planning with an eye on eliciting high-level student mathematical discourse.

---

[2]The 2006 OMLI discourse data suggest that most of the student mathematical discourse in K–12 classrooms is what we have characterized as low-level mathematical discourse with few, if any, instances of high-level discourse (Weaver & Dick, 2006).

Drawing upon the information from the collaborative planning (i.e., the analysis of the lessons with regard to the cognitive demand of the tasks and the potential for student mathematical discourse) the researcher examined the relationships between the intended and enacted curriculum as this work yielded evidence of the mathematical reasoning students engaged in at early points in the unit. Since the data was gathered across an investigation, it was also possible to examine if it took more than one class session for a teacher to draw out higher levels of mathematical discourse. In collaboration with the researcher, the teachers worked to identify potential places to raise the level of mathematical discourse and then attempted to draw out higher levels of mathematical discourse from students than might have otherwise come forward if the curriculum were implemented as written.

## The Three Contexts

Ms. Johnson[3] and Ms. Loring are veteran elementary teachers. Their classrooms share a portable wall and they often plan together. These fourth-grade teachers loop with their students as they move from fourth to fifth grade. They are in the their third year using the *Investigations* (Mokros & Russell, 1995) and the second time teaching the unit *Money, Miles, and Large Numbers* due to their looping.

Ms. Rand is a veteran elementary teacher and a teacher leader in her district. While she does not have grade-level colleagues with whom she works regularly, at the time of this study she was mentoring a student teacher. She is a third-grade teacher enacting the *Fair Shares* for the third time.

Ms. Anders, also a veteran teacher and teacher leader in her district, teaches a fourth/fifth grade split. Like Ms. Johnson and Ms. Loring, she shares a wall and students with another teacher. For mathematics, the teachers worked with their students by grade level. Ms. Anders worked with the fourth graders while her colleague worked with the fifth graders. Ms. Anders taught math using *Investigations* (Mokros & Russell, 1995; Russell & Economopoulos, 2008) for three years prior to this study and was field-testing the new edition of *Landmarks and Large Numbers* at the time of this collaborative work.

Each teacher's classroom was physically structured so that students were seated in groups of four. Students were commonly directed to work privately when a task was first posed and then later moved to collaborating with a partner or with their small group. The nature of the whole group discussion (i.e., who participates, mode of mathematical discourse[4]) showed

---

[3]As per my agreement with these teachers, the names used for them and their students are pseudonyms.
[4]The OMLI Classroom Observation Tool (Weaver & Dick, 2006) defines discourse modes that refer to who the student addresses during the discourse. Four modes were identified: student to teacher, student to student, student to group (small or whole), and student to self (via private reflection in a journal, for example).

some variance from one classroom to the next, however these variations are not discussed in this paper as the main focus is on the levels of the student mathematical discourse.

Each of these teachers is dedicated to *Standards*-based (NCTM, 2000) instructional practice and had experience using reform-oriented curriculum prior to their districts' adoption of the *Investigations* (Mokros & Russell, 1995; Russell & Economopoulos, 2008) curriculum. Participation in the OMLI project was an opportunity for them to deepen their mathematical content knowledge, their leadership and facilitation skills, and their classroom practices to support each student's mathematics learning.

*Data Collection and Analysis*

Using a continuum of types of student mathematical discourse (see Table 1), the researcher inferred the levels of student mathematical discourse as prompted by the written curriculum materials (the intended curriculum) and assessed the levels of discourse evident in videotapes of third and fourth grade lessons (the enacted curriculum). The lessons chosen for analysis were, with the exception of Ms. Anders' classroom, from the first or second lesson from the focus unit. The purpose of selecting this "early" lesson was to respond to the question/concern stated previously that it was not until the second or third day of a lesson or until the end of a unit that students would engage in higher-level discourse.

The team, teacher(s) and researcher, used the levels of discourse and the *Thinking Through a Lesson Protocol* (Stein, et. al., under consideration) to inform the collaborative planning. Prior to the lesson enactment, the team would meet and discuss the mathematical ideas of the lesson; anticipate students' responses to mathematical task(s); infer the level of student discourse that the task is likely to elicit; and as appropriate, "tweak" the task(s) so that higher-level thinking and discourse would emerge during small and whole group discussion. These lessons were videotaped and analyzed to determine the extent to which the team was successful in promoting higher-level student mathematical discourse including relating, predicting, conjecturing, (type 4) justifying, and generalizing (type 5).

## RESULTS AND DISCUSSION

In the unit *Money, Miles, and Large Numbers* the team recognized that many of the tasks, as written, would yield only low levels of student mathematical discourse (i.e., answering, stating, sharing, explaining). (See Table 2 for details regarding the level of student mathematical discourse in both the intended curriculum and selected excerpts from the enacted curriculum in Ms. Johnson's and Ms. Loring's classrooms.) The team used this information to inform the set-up and implementation of the tasks. For *Ways to Make a Dollar,* instead of simply focusing on ways buy two items and spend exactly a dollar we decided to take advantage of the extension activity of determining all the possible ways to buy two items and spend exactly a dollar as that would deepen the discourse to relating (as students are pressed to continue their lists in an organization fashion), and predicting and justifying the total number of ways to make a dollar.

A second component of this lesson was a game called Close to 100. Again, the levels of discourse prompted by the curriculum materials as written were low level. The team discussed plans for enhancing the discourse and determined that asking students to talk about their strategies would be an important first step; and for higher-level discourse we would want students to articulate "general" strategies for playing the game. Examples of these are listed in Table 2, Type 5; one student in Ms. Loring's class commented, "it really depends on the cards I have, which strategy I use." Two strategies included making two values close to 50 or making 90 with the digits in the tens column and ten with the digits in the ones. The critical analysis of deciding which strategy based on the cards was moving students toward the kind of thinking we wanted them to employ. We later asked them to consider, to what extent these strategies would work for *Close to 1000* and beyond.

For the unit *Fair Shares,* Ms. Rand and I found that there were opportunities across the unit as written for all levels of mathematical discourse (see Table 3). The focus of our collaborative planning was on supporting students as they considered all the different solutions that were presented, knowing that fractions can be a difficult topic. For *Sharing Seven Brownies* the actual student discourse was split between explanation and justification. The paper brownies were important tools for students both as they solved the problem and as they communicated and represented their mathematical thinking. Two aspects that students felt were important to justify in their approaches were that they had used all seven brownies and that all the shares were "fair shares." This may have been due in part to the fact that in the previous lesson, as students were making fraction cards, some did not create equal size pieces either due to rough folding or imprecise cutting, so they would trim and then their pieces were not of "standard" size and could not be used.

The lesson continued with *More Sharing Problems.* Ms. Rand and I felt that with appropriate questioning, time, and plenty of paper brownies students could see that multiple correct solutions to the problems are equivalent[5]. For example, two solutions were presented for three brownies divided between two people: Kiara's solution of 6/4 and Alec's solution of 3/2. Given

---

[5]This conjecture was based on the observation we had made as students were listing Fraction Facts and students were eager to prove equivalence through a visual model of layering fraction pieces on top of other pieces.

**TABLE 2. Discourse Data by Type—Money, Miles, and Large Numbers**

| Discourse Types | Intended Curriculum | Enacted Curriculum—Ms. Johnson | Enacted Curriculum—Ms. Loring |
|---|---|---|---|
| Type 1: Answering, Stating, or Sharing | **Ways to Make a Dollar**<br>A student suggests a way to buy two items and spend exactly $1.00 – such as 50¢ and 50¢; 30¢ and 70¢.<br><br>**Close to 100**<br>A student shares the results after constructing two two-digit numbers with the sum as close as possible to 100. | **Ways to Make a Dollar**<br>• "How about 50¢ and 50¢?"<br>• "We had 30¢ plus 70¢."<br>• "90¢ and 10¢"<br>• "95¢ and 5¢"<br>• "99¢ and 1¢"<br>• "$1.00 and 0"<br><br>**Close to 100**<br>62 and 42… I got 104. | **Ways to Make a Dollar**<br>How many different ways do you think there is to make $1.00?<br>• "100"<br>• "a lot of ways"<br>• "…probably about 50"<br><br>**Close to 100**<br>• "I know what I would do. I would do 80 and 21. That's 101."<br>• "I'd use 98 and 02." |
| Type 2: Explaining | **Ways to Make a Dollar**<br>A student shares what he or she did to find combinations to make $1.00. | **Ways to Make a Dollar**<br>• "4 + 6 is 10¢ and 70 and 20 is 90¢ and 90¢ and 10¢ is $1.00."<br>• "You take the 4 from 24 and put it on the 76 to make 80, then you take the 20 and add it to the 80 and it makes $1.00."<br><br>**Close to 100**<br>• "I had 26 and 74 to get 100. I added the 4 and 6 and got 10. The I added the 7 and 2 which is 90 plus 10 is 100."<br>• "I used the Wild Card as 0, so this is 90 [pointing to 90 + 0], I used the 5 as just a 5, so I am up to 95 then I added the 4 to it, I'm up to 99."<br>• "First I took the 3 and the Wild Card I made a 7 because that made 10, the I took this 5 and 4 to make 90 and 90 and 10 make 100." | **Ways to Make a Dollar**<br>• "One way that I could to this would be to list all of the ways to make 100, but that would take a long time. I'm trying to think of a faster way."<br>• "Maybe we could use a hundreds chart…. (student places fingers on the chart pairing 99 and 1, 98 and 2, 97 and 3 and counts the pairs)"<br><br>**Close to 100**<br>• "My strategy was picking numbers that were the lowest number and the highest number like 20 and 70 to get 90 and then I'd make it like 23 and 75, so it would be 98… I was looking at the tens and then I would look at the ones and add that to it."<br>• "I would choose numbers that are close to the middle… but that doesn't work very well with this hand of cards." |

**Type 3: Questioning or Challenging**

**Ways to Make a Dollar**
- "Can we use $1.00 and 0 as a way to make a dollar?"

**Ways to Make a Dollar**
- "I'm just not sure that there is really 50 ways. I know they will repeat again after 50 but I don't know if there are 50."

**Type 4: Relating, Predicting, or Conjecturing**

**Ways to Make a Dollar**
A student notices and uses the relationships among the combinations to generate other combinations.[1]

**Ways to Make a Dollar**
I noticed that one person in your group would say something and then another person would use that idea...
- "Yea, like Addie said 74 and 26 after I said 75 and 25"
- "We started by going with higher number and then we went to lower numbers like 1 and 99, 2 and 98, 3 and 97..."

Does anyone have a prediction about how many different ways there are to make a dollar when buying just two items?
- "I think there could be like 100 ways because there is like 1 + 99, 2 + 98... and we can go all the way to like 100."

**Ways to Make a Dollar**
How many different ways do you think there is to make $1.00?
- "There's 100 cents in a dollar, so there's probably 100 ways."

*(continued)*

**TABLE 2. (cont.)**

| Discourse Types | Intended Curriculum | Enacted Curriculum—Ms. Johnson | Enacted Curriculum—Ms. Loring |
|---|---|---|---|
| Type 5: Justifying or Generalizing | | **Ways to Make a Dollar** "We need to buy two items and it equals a dollar. One of the items could be free. It's like buy this candy and get one free." <br>• "I think there is only 50 ways because when you are adding them together you are using 2 numbers." <br>• "I think there is 100 ways because there is 100 cents in a dollar and you can have $1 + 99$ and $99 + 1$." <br>**Close to 100** <br>• "... I added up the ones to see how much tens I would need first. Since I had 9 [ones] I decided to use 90 because I knew it would be one away from 100." | **Ways to Make a Dollar** <br>• "If you did 100, you'd do them twice because you'd be switching the numbers around. So really you've counted them twice, it would be half of 100 or 50 ways." <br>**Close to 100** <br>• "First I work on the tens and try to make 100 then I go to the ones and try to make something small so that I don't go too far over. Otherwise I try to make 90 in the tens 10 in the ones." |

[1]The teacher is not prompted to ask students about their strategy to generate combinations they are simply prompted to notice if students made a random list or if combinations are related.

**Table 3. Discourse Data by Type—Fair Shares**

| Discourse Types | Intended Curriculum | Enacted Curriculum: Ms. Rand |
|---|---|---|
| Type 1: Answering, Stating, or Sharing | **Sharing Seven Brownies**<br>A student provides $1 + \frac{1}{2} + \frac{1}{4}$ as a solution to the task. | **Sharing Seven Brownies**<br>• "Each person gets a whole brownie and 3 pieces."<br>• "It's $\frac{1}{4}$ because there are 4 equal parts." |
| Type 2: Explaining | **Sharing Seven Brownies**<br>A student describes the model he or she created, but does not explain why. | **Sharing Seven Brownies**<br>• "I cut the extra brownies into 4 little pieces... I started with giving the big pieces (brownies) out and there were 3 extra brownies... then we cut the 3 extra brownies into 4 equal pieces and gave them out."<br>• "I folded all 7 brownies into 4 pieces and then I cutted (sic) them and shared them. (Student shows sharing the pieces among the 4 people)."<br>• "First I cut all the brownies in half and I shared them, but then I had to take the last one and cut it into fourths... each gets $\frac{1}{2} + \frac{1}{2} + \frac{1}{4}$."<br>**More Sharing Problems**<br>• "I took all of my brownies and I folded them in half and then cut them in half and I know all of them are equal... they each got 3/2." |
| Type 3: Questioning or Challenging | **Sharing Seven Brownies**<br>A student wonders if 7/4 is the same share as $1 + \frac{1}{2} + \frac{1}{4}$. | **Sharing Seven Brownies**<br>"Can I cut them in half?" |

*(continued)*

**Table 3.  (cont.)**

| Discourse Types | Intended Curriculum | Enacted Curriculum: Ms. Rand |
|---|---|---|
| Type 4: Relating, Predicting, or Conjecturing | **Sharing Seven Brownies**<br>A student predicts that each person will get between one and two brownies since there are more than 4 and less than 8 brownies to start.<br>**More Sharing Problems**<br>A student says that the solution 3/2 is the same as ½ + ½ + ½ because the 3 is counting the number of pieces of one half. | **Sharing Seven Brownies**<br>Are these answers the same? "<br>• They are… They all started with 7 brownies.<br>**More Sharing Problems**<br>Kiara says the answer is 6/4 and Alec says the answer is 3/2, do you see any connection there?<br>• "Kiara almost did the same thing as me [he displays his paper with 3/2 and says] she just cut hers in half."<br>• "One-and-a-half is the same because two-halves together make one whole with the other half it is the same [shows on the paper how this division of the whole shows the same three halves." |
| Type 5: Justifying or Generalizing | **Sharing Seven Brownies**<br>A student shows that each person receives the same amount of brownie by stacking "brownie" shares to show equivalence. | **Sharing Seven Brownies**<br>• A student records the following number statement to show that her method used all seven brownies:<br>• $1 + 1 + 1 + 1 + ½ + ½ + ½ + ½ + ¼ + ¼ + ¼ + ¼ = 7$<br>• "Everyone has a fair share because their pieces are the same [shows by stacking]."<br>• "It's 7/4 'cuz ¼ is one of the little brownies and the top changes— that is how many little brownies each person gets—like 10/8 yesterday." |

**TABLE 4. Discourse Data by Type—Landmarks and Large Numbers**

| Discourse Types | Intended Curriculum | Enacted Curriculum—Ms. Anders |
|---|---|---|
| Type 1: Answering, Stating, or Sharing | **What's Happening**<br>A student states that they need to drive 315 miles in the second day. | **What's Happening**<br>What are we trying to figure out in this problem?<br>• "How far away they are from Grandma's house" |
| Type 2: Explaining | **What's Happening**<br>A student draws a number line and shows the jumps from 319 to 634 (e.g., first a jump of 1 to 320, then a jump of 80 to 400, then a jump of 200 to 600, and finally a jump of 34 to 634; next the student adds all the jumps together).<br><br>**Subtraction Story Problems**<br>A student draws a sketch to compare the two distances traveled and see how much farther Ms. Santos drove. | **What's Happening**<br>• "I go 1 mile to get from 319 to 320. I go 314 miles to get from 320 to 633. Does that make sense? So I go 315 miles to get from 319 to 634. Does that make sense? ... because I added those together (points to 1 and 314)."<br>• "319 plus 81 is 400; (be thinking about why Tyler would take each of these steps, why would he do it that way?) 400 plus 234 is 634 and 81 plus 234 is 315."<br>• "First I added 300 onto 319. That got me to 619. Then I did plus 15 and that got me to 634. Then I added 300 and 15 and got 315."<br>• "I did 34 minus 19 and that was 15. And then I did 600 minus 300 and that's 300. So I plussed 300 and 15 together and got 315."<br><br>**Subtraction Story Problems**<br>• "If you start with that number line then all you need to do is plus the things that you added on to get the answer."<br>• "You could do minusing too. You could do 1300 and keep minusing until you get to 446 and then plus those together. That would be the same answer."<br><br>*(continued)* |

**TABLE 4. (cont.)**

| Discourse Types | Intended Curriculum | Enacted Curriculum—Ms. Anders |
|---|---|---|
| Type 3: Questioning or Challenging | **What's Happening**<br>A student asks why another student made a jump of 80 to 400. | **What's Happening**<br>• "I don't really get it, could you explain it again so that I can understand."<br>• "If you are adding on that amount, don't you need to subtract it?"<br>• "Where did you get the 5 from? (asking about a compensation strategy)" |
| Type 4: Relating, Predicting, or Conjecturing | | **What's Happening**<br>What did Tyler and Katie both do to solve this?<br>• "...added on to 319 because they drove that much by the first day."<br>• "...stopped at 634 [Grandmother's House]."<br>Connecting a number line to numeric methods<br>• "You'd start at 319 and hop 81 to 400. Then you'd hop 234 to get to 634.... The hops are how many miles are left." |
| Type 5: Justifying or Generalizing | **What's Happening**<br>A student provides the number line explanation shown above (in type 2) but in addition to stating the process also explains why these steps were taken (e.g., I jump 80 from 320 because 400 is an easier number to work with). | **What's Happening**<br>• (In response to a question from a classmate) "I did 319 plus 81 because I knew that would put me on an even number, like a whole hundred. Those are a lot easier to work with. So I got 400, then plus the 234 because that would make the exact miles I needed to go, 634. The two distances that I added on, 81 and 234 were what I add to get my answer." |

the question: Kiara says the answer is 6/4 and Alec says the answer is 3/2, do you see any connection there? students explored and began making connections between these and other solutions. This search for equivalent solutions to problems continued throughout the unit indicating that students were developing a habit of mind of looking beyond a solution and using representations to justify what they were seeing. This was not a significant change from the lesson as written, however it combined goals of several smaller lessons to allow depth over breadth.

Ms. Anders and I had the opportunity to plan from the new edition a fourth-grade *Investigations* (Russell & Economopoulos, 2008) unit called *Landmarks and Large Numbers*. The updated version of the materials provides extensive support for the teacher in terms of deepening her understanding of content and what to look for in student thinking. For students there are additional tools to support sense-making and there is a clear expectation for communication about reasoning rather than simply reporting.

While the work in Ms. Anders' classroom occurred mid-unit, it represents the students' first experiences with solving three-digit subtraction problems and representing subtraction on a number line. The lesson as written included opportunities for students to both explain and justify their mathematical thinking (see Table 4). During the whole group discussion of *What's Happening*, as each student shared, Ms. Anders prompted students to think about *why* that student might be taking each of the steps he/she took to solve the problem. Also notable, were the types of questions that students asked one another (see Table 4, Type 3), as it was clear that they were listening to understand one another's thinking. Even though the initial explanation Tyler provided for his method did not include his reasoning, when challenged by a classmate who did not understand, he was able to articulate a rationale for each move related to the numbers he chose and how these numbers made his calculations easier for him.

## CONCLUSION

This collaborative study gave the opportunity to study issues related to the intended and the enacted curriculum as well as consider what is possible from students as they are learning new content. As seen in the previous section, a range of levels of discourse is supported in the curriculum as written, for most units.[6] While it may be the case for some units, that higher-level discourse does not occur until later in the unit, it is prompted by the curriculum studied here. With regard to what students could do, there was no

---

[6]As there were fewer opportunities for higher-level discourse in the *Money, Miles, and Large Numbers* unit.

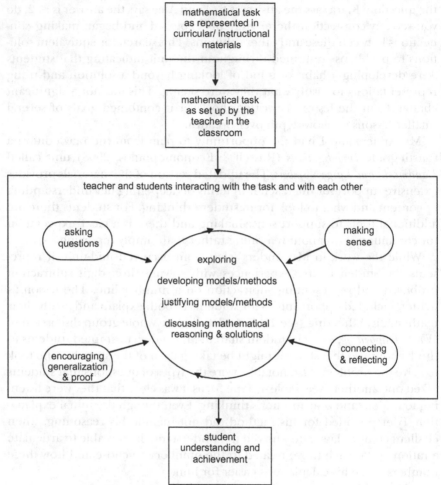

**FIGURE 2.** **Relationships among the task, the teacher and student interactions with the task and each other, and the students' understanding and achievement. Adapted from Stein, Grover, & Henningsen, 1996 and Rigelman, 2003.**

question that with the appropriate time and intellectual support (questions, norms) these third and fourth graders engaged in high level discourse.

In Figure 1, I offer the *Mathematics Tasks Framework* (Stein, Grover, & Henningsen, 1996) as a means of thinking about the interactions between the task/curriculum, teacher, and students. In this section I offer an expanded framework that provides more detail for the implementation phase (see Figure 2). The data presented in this paper and elsewhere (Rigelman,

2003) suggest that students in reform-oriented mathematics classrooms regularly engage in:

1. exploring;
2. developing models/methods;
3. justifying models/methods; and
4. discussing mathematical reasoning and solutions.

Further, as teachers and students in these classrooms interact with one another and with the task/curriculum they deepen the mathematical work within their community by:

1. asking questions;
2. making sense;
3. connecting and reflecting; and
4. encouraging generalization and proof.

As this mathematical work deepens, student understanding deepens and achievement improves as well (Hiebert, Carpenter, Fennema, Fuson, Wearne, & Murray, 1997; Henningsen & Stein, 1997; Stein, Smith, Henningsen, & Silver, 2000).

## REFERENCES

Battista, M. T. (2007). Learning with understanding: Principles and processes in the construction of meaning for geometric ideas. In Martin, W. G., Strutchens, M. E., & Elliott, P. C. (Eds.), *The learning of mathematics* (pp. 65–99). Reston, VA: National Council of Teachers of Mathematics.

Drake, C., & Sherin, M. G. (2006). Practicing change: Curriculum adaptation and teacher narrative in the context of mathematics education reform. *Curriculum Inquiry, 36*(2), 153–187

Henningsen, M., & Stein, M. K. (1997). Mathematical tasks and student cognition: Classroom-based factors that support and inhibit high-level mathematical thinking and reasoning. *Journal of Research in Mathematics Education, 28*(5), 524–549.

Hiebert, J., Carpenter, T. P., Fennema, E., Fuson, K. C., Wearne, D., & Murray, H. (1997). *Making sense: Teaching and learning mathematics with understanding.* Portsmouth, NH: Heinemann.

Hill, H. C., Schilling, S. G., & Ball, D. L. (2004). Developing measures of teachers' mathematics knowledge for teaching. *Elementary School Journal, 105*, 11–30.

Hufford-Ackles, K., Fuson, K. C., & Sherin, M. G. (2004). Describing levels and components of a math-talk learning community. *Journal for Research in Mathematics Education, 35*(2), 81–116.

Hughes, E. K., & Smith, M. S. (2004). *Thinking through a lesson: Lesson planning as evidence of and a vehicle for teacher learning.* Poster presented as part of a symposium, "Developing a Knowledge Base for Teaching: Learning Content and Pedagogy in a Course on Patterns and Functions" at the annual meeting of the American Educational Research Association, San Diego, CA.

National Council of Teachers of Mathematics. (2000). *Principles and standards for school mathematics.* Reston, Virginia: Author.

National Council of Teachers of Mathematics. (2007). *Mathematics teaching today.* Reston, Virginia: Author.

Mokros, J., & Russell, S. J. (1995). *Investigations in number, data, and space.* Palo Alto: Dale Seymour Publications.

Remillard, J. T. (1999). Curriculum materials in mathematics education reform: A framework for examining teachers' curriculum development. *Curriculum Inquiry, 29*(3), 315–342.

Rigelman, N. R. M. (2003). *Teaching mathematical problem solving in the context of Oregon's educational reform.* Unpublished doctoral dissertation, Portland State University, 2002. *Dissertation Abstracts International* 63-06: 2169.

Russell, S. J. & Economopoulos, K. (2008). *Investigations in number, data, and space* (2nd ed.). Parsippany, NJ: Pearson Learning Group.

Sherin, M. G. (2002). A balancing act: Developing a discourse community in a mathematics classroom. *Journal of Mathematics Teacher Education, 5*, 205–233.

Smith, M. S., & Stein, M. K. (1998). Selecting and creating mathematical tasks: From research to practice. *Mathematics Teaching in the Middle School, 3*(5), 344–350.

Stein, M. K., Grover, B. W., & Henningsen, M. (1996). Building student capacity for mathematical thinking and reasoning: An analysis of mathematical tasks used in reform classrooms. *American Educational Research Journal, 33*(2), 455–488.

Stein, M. K., Smith, M. S., Henningsen, M. A., & Silver, E. A. (2000). *Implementing standards-based mathematical instruction: A casebook for professional development.* New York: Teachers College Press.

Stein, M. K., Engle, R. A., Hughes, E. K. & Smith, M. S. (under consideration). *Orchestrating productive mathematical discussions: Helping teachers learn to better incorporate student thinking.*

Weaver, D., & Dick, T. (September 2006). *Assessing the Quantity and Quality of Student Discourse in Mathematics Classrooms, Year 1 Results.* Paper presented at Math Science Partnership Evaluation Summit II, Minneapolis, MN. Available at: http://ormath.mspnet.org/index.cfm/14122.

# CHAPTER 9

# BEYOND TACIT LANGUAGE CHOICE TO PURPOSEFUL DISCOURSE PRACTICES

**Beth Herbel-Eisenmann**

In this chapter, I describe some of the primary activities that took place as part of a five-year professional-development experience. One long-term goal of the project was to better understand how having middle-school mathematics teachers do action research on their discourse practices might impact their beliefs and practice over time. The specific project activities I describe here include the ways in which we: a) read analyses of the baseline data; b) created mappings of "what was closest to my heart in my teaching"; c) participated in a study group on classroom discourse, mathematics classroom discourse, and action research; and d) identified a performance gap. I do not make claims about what these activities have done *per se*, but rather focus on describing the activities themselves in order to make them available to others interested in working with mathematics teachers on classroom discourse.

**KEYWORDS:** language choice, action research, study groups, teacher beliefs, professional development

## INTRODUCTION

As a beginning mathematics teacher, my "model" for what to do with language consisted of writing vocabulary words on the board, followed by recording detailed definitions. Students were then asked to commit the

*The Role of Mathematics Discourse in Producing Leaders of Discourse*, pages 173–198

vocabulary words and definitions to memory and use them appropriately in class and on their homework, quizzes, and tests. This model is what I observed for my entire mathematics education and my teacher preparation program. When I started graduate school, I soon realized that there was more to be concerned with related to language than simply learning vocabulary and definitions. I came to understand that there were at least four reasons why classroom discourse was something that needed more careful and thoughtful attention in mathematics education:

1. mathematics is a specialized form of literacy;
2. spoken language is a primary mode of teaching/learning;
3. the particular context in which language is used plays a role in what is appropriate to say and do; and
4. language is intimately related to culture and identity (see Herbel-Eisenmann, Cirillo, & Skowronski, 2009, for further articulation of these reasons).

As I continued to read empirical studies on classroom discourse, I became discontented with the expectations mathematics education researchers seemed to have of teachers. That is, I noticed that ideas from discourse analysis seemed to be used to illustrate how teachers were *not* meeting the vision of the *Standards* (NCTM, 1989, 1991, 2000) documents, rather than being used to better understand what teachers were doing and possibly considering what other things they might have done. Also, they often did not articulate how we, as teacher educators, might better work with other teachers around issues of classroom discourse. This led me to collaborate with a group of eight secondary mathematics teachers to better understand what they might do with ideas from classroom discourse and mathematics classroom discourse literature, given the opportunity to work together over a substantial period of time, reflecting on their practice and implementing action research projects related to the new ideas they were encountering. Over the past four years we have engaged in a series of learning activities together. As we have neared the end of our work, the teacher-researchers (TRs) have talked to us (myself and the graduate students who work with us) about which aspects of the project were most important to their learning about and changing their classroom discourse. I will try not to make claims about what these activities have done *per se*, but rather will focus on describing the activities themselves.

## THE PROJECT AND PEOPLE

The broader goals and phases of the project are to examine the following:

- the nature of mathematical discourse in middle school mathematics classrooms;
- the ways in which middle school mathematics teachers' beliefs impact their discourse when working to enact reform-oriented instruction; and
- how this information can be used to incorporate action research using concepts and tools of discourse analysis to improve mathematics instruction.

The project itself was organized in four phases:

1. recruiting partner teachers, engaging in conceptual work, and preparing the new research assistants for the upcoming work;
2. collecting baseline data (e.g., interviews; four weeks of classroom observations that were videotaped and transcribed; teacher-authored week plans, daily reflections, and weekly reflections) about the teacher-researchers' classroom discourse;
3. organizing study group meetings that involved reading literature on classroom discourse, mathematics classroom discourse, and action research; and
4. supporting teacher-researchers through cycles of action research.

Throughout all four phases, we had project meetings that ranged from three hours to full-day and overnight retreats and that involved doing activities together (e.g. examining mathematical tasks using the QUASAR task framework [Stein, Smith, Henningsen, & Silver, 2000]), discussing readings, watching video-tapes of other teachers (e.g., Deborah Ball's "Shea Numbers"), watching video-tapes of each other, and so on.

The people involved in the project included myself, a university researcher (UR); two graduate research assistants (also URs); and eight middle-grades (grades 6–10) mathematics teacher TRs from seven schools in a Midwestern state in the United States. The TRs in the group taught in different kinds of communities with students from varying levels of poverty (free and reduced lunch percentages varied from 12 to over 65 percent) and in different kinds of schools, including some high schools (HS, grades 9–12) and some middle schools (MS, grades 6–8). Three of the TRs were working in schools where NSF-funded[1] curriculum materials had been used for more than 10 years. One of these TRs was also a pilot teacher for the NSF-funded curriculum materials designed to provide cognitively-demanding mathematical tasks, and encourage teachers to engage students in mathematical discussions, explanations, and justifications about big mathemati-

[1]NSF-funded curriculum materials are those that were developed with funding from NSF in order to embody the ideals put forth by the National Council of Teachers of Mathematics (NCTM, 1989).

**TABLE 1. Description of Teacher-Researchers**

| TR | Grade | School Setting | Certification | Yrs Teaching | Curriculum Materials |
|----|-------|----------------|---------------|--------------|----------------------|
| Tonya | 6 | Rural, MS | Elem | 21 | NSF |
| Dave | 6 | Urban, MS | Elem | 7 | Trad |
| Annie | 7 | Rural, MS | Elem/MAT*** | 17 | NSF |
| Penny | 8 | Urban, Title I*, MS | Sec | 18 | Trad |
| Jackie | 8 | Suburban, MS | Sec/MS*** | 14 | NSF |
| Lisa | 8 | Urban, Gifted**, HS | Sec | 9 | Trad |
| Jim | 8 | Urban, MS | Sec/MSM*** | 14 | Trad |
| Matt | 10 | Suburban, HS | Sec/MAT | 2 | Trad |

* A Title I school has a high percentage of students living in poverty (at least 40%) and is therefore provided with additional financial assistance from the government.

** Students in this school scored in the 95th percentile on standardized tests. Students in eighth grade math in this school might actually be in sixth or seventh grade

*** MAT is a Masters of Arts in Teaching; MS is a Masters of Science; MSM is Masters of School Mathematics

cal ideas.[2] The other five TRs taught in schools where more conventional curriculum materials were adopted and used. See Table 1 for a summary of each teacher's background and setting.

The TRs in these classrooms were purposefully selected to vary gender, context of teaching situation, certification level, years of teaching experience, extent of involvement in professional development, and reasons for entering the teaching profession. Using diverse sites and multiple participants is one way of increasing the level of trust in findings related to the impact of action research on teachers' practices (Doerr & Tinto, 2000). The number of years they had been teaching mathematics ranged from 2 to 18 years. All of the TRs participated in district-mandated professional development outside this project. About half of these TRs attended state conferences for mathematics teachers and two were active in NCTM at the regional and national levels. Only one of the TRs was unfamiliar with the NCTM prior to the project. Two of the TRs were National Board Certified[3]

[2]For specific distinctions between NSF-funded curriculum materials and more conventional materials in relationship to algebra, see Star, Herbel-Eisenmann, and Smith (1999).
[3]In the United States, teachers can apply for National Board Certification whereby they show that they have met rigorous national standards set by The National Board for Professional Teaching Standards through intensive study, expert evaluation, self-assessment, and peer review.

and two were awarded the Presidential Award for Excellence in Mathematics and Science Teaching, the highest national award given in the United States for mathematics teaching. All of the TRs volunteered to be involved in the project and stated that their purpose for becoming involved was related to continuing to improve their teaching.

## RELATED LITERATURE

There is evidence that discourse practices in mathematics classrooms have not changed much in the last two decades (Spillane & Zeuli, 1999). Research has shown that mathematics teachers' discourse patterns are quite traditional (e.g., Stigler & Hiebert, 1999), including those of teachers who are attempting to change their classroom practices (e.g., Herbel-Eisenmann, Lubienski, & Id-Deen, 2006). These findings are important given that the reform movement in mathematics education has made some particular demands on teachers that have implications for classroom discourse. The *Standards* aim to promote conceptual understanding and sense making instead of the procedural emphasis that often takes precedence in more traditional mathematics teaching. Classroom discourse can affect the learning environment and student engagement in learning mathematics in the ways proposed by the *Standards*, but only if some of the discourse patterns in mathematics instruction are changed from a transmission model of communication to an inquiry-based one. The value of using methods of discourse analysis in mathematics classrooms is that they can "document the difficulties of implementing genuine instructional reform in classrooms" (Forman, Larreamendy-Joerns, Stein, & Brown, 1998, p. 334). In order to achieve goals for teaching mathematics that embrace unconventional patterns of discourse, teachers need to make explicit and examine the beliefs and conceptions they bring from their own experience as students of mathematics. As Thompson (1992) stated:

> The tendency of teachers to *interpret new ideas and techniques* through *old mindsets*—even when the ideas have been enthusiastically embraced—should alert us against measuring the fruitfulness of our work in superficial ways (p. 143, emphasis added).

Even when teachers' professed beliefs are aligned with the ideals put forth by NCTM, the *mindsets* they have inherited from participation in more conventional mathematics classrooms are implicitly embedded in and carried by the language teachers use. This may directly influence the norms and discourse that teachers negotiate in their classrooms. Yet, in order to examine language choice, a teacher would need to be consciously and explicitly aware of the choices s/he is making. Researchers have reported that teachers[4] rarely have this level of awareness (Barnes, 1969), a finding that was more recently reported in mathematics education (Herbel-Eisenmann, 2000). That said, secondary mathematics teachers can and should be gathering evidence of embedded beliefs from their own classrooms' discourse practices, thereby determining what counts as evidence. Having specific artifacts from their own classrooms is a powerful form of evidence that can help them to better align their professed beliefs and their practices. By examining the prevalent forms and functions of the classroom communication system, the norms become more apparent. These norms can be examined to see how a teacher might be undermining or promoting the kind of discourse s/he may want to establish. Discourse analysis shows a key linkage between beliefs and practices. In this project, secondary mathematics TRs were offered concepts and tools of discourse analysis as a lens through which to view and reflect on their teaching.

## LEARNING ABOUT, REFLECTING ON, AND CHANGING CLASSROOM DISCOURSE

In this section, I describe some of the activities that the TRs identified as important to their learning about, reflecting on and changing their classroom discourse:

1. reading analyses of the baseline data;
2. creating mappings of "what was closest to my heart in my teaching";
3. participating in a study group on classroom discourse, mathematics classroom discourse, and action research; and
4. identifying a performance gap.

All four of these activities were important to the ways in which the TRs developed their action research focus. If the reader is interested in more detailed descriptions of the action research projects, see Herbel-Eisenmann and Cirillo (2009).

### Reading Analyses of Baseline Data

Each TR allowed us to observe a typical week of his/her teaching four times (about every other month) during the 2005–06 school year. These observations were videotaped and audio-taped and one university-research-

---

[4]We want to be clear that we are not pointing fingers at teachers when we say this. In fact, we believe that people, in general, are not very aware of their discourse patterns primarily because we learn discourse practices *in* the practice.

er took *in-situ* field-notes and ran the video-camera. The TRs wore micro-phones and an additional microphone was attached to the camera and placed in the middle of the classroom. This typically meant that we were able to get all of the TR's utterances recorded, but were less successful re-cording all of the student utterances. All of these classroom observations were transcribed and imported into Transana (Fassnacht & Woods, 2005), a qualitative database management system. At the end of the 2005–06 school year, we distributed the Teacher's Communication Behavior Questionnaire (TCBQ, see She & Fisher, 2000) to each teacher and to the students in his/her focus class. The TCBQ is a survey instrument that was developed to assess student perceptions of the following five teacher communication behaviors in science classrooms:

1. challenging;
2. encouragement and praise;
3. non-verbal support;
4. understanding and friendly; and
5. controlling.

We did some descriptive statistics based on this questionnaire and wrote a brief report of how the teacher's responses compared to their students (see Cirillo & Herbel-Eisenmann, 2006 for more about this analysis). After col-lecting this data, we did some quantitative and qualitative analyses to share with the teachers. Because there is a paucity of this kind of reciprocal work with practicing teachers available, we were unsure what information might be interesting or useful to the mathematics teachers.

After reading a range of literature on mathematics classroom discourse, we were convinced that a combination of systemic functional grammar and critical discourse analysis would help us to better understand how the teach-ers construed mathematics through language as well as to examine issues related to authority and control[5] (issues that the TRs discussed when they watched themselves on video). Using modified definitions of "activity struc-tures" (based on the work of Lemke, 1990), our first level of coding con-sisted of creating timelines of the various activity structures that occurred. For example, we created collections in Transana of the teacher's "going over homework," "small group work," "seatwork," and "whole class work." These collections allowed us to create pie charts of the percent time spent on each of these kinds of activities, both for each week we observed as well as across the four weeks of observation (see Figure 1 for an example of the pie charts). We also used a computer program developed by one of the un-

---

[5]For some of our analyses involving authority, control, and positioning, see Herbel-Eisenmann (2009), Herbel-Eisenmann, Wagner, and Cortes (July, 2008), Wagner and Herbel-Eisenmann (2008), and Wagner and Herbel-Eisenmann (in press).

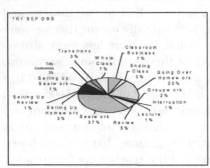
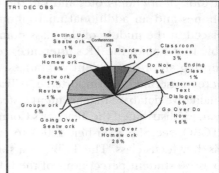
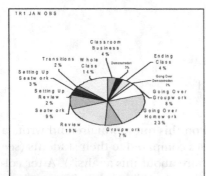
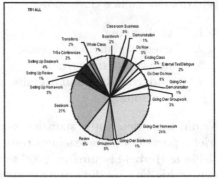

**FIGURE 1. Sample Pie Charts of Percent Time Spent on Activity Structures**

dergraduate research assistants to do a turn-length word count, comparing the number of words spoken by each teacher each week to the number of words spoken by the students in the classroom (see Figure 2 for an example of the box and stem plots we made).

In August 2006, we met with the TRs and gave them each a binder with the above information included in it. We did a short presentation explaining the various parts of the binder (i.e., the pie charts with time spent on the various activity structures, the turn-length analysis, and the TCBQ analysis) and then gave the TRs time to read the information in their binders. We had a brief discussion with them and then asked them to read the information more carefully so that we could discuss it again at our next meeting. In the meantime, we asked them to decide which of the activity structures they would like to know more about in order to help us to decide where to focus our detailed qualitative analysis of their classroom discourse.

After each TR selected his/her activity structures of interest, we analyzed the data, drawing on literature from systemic functional grammar (e.g., Halliday, 1978; Lemke, 1990; Pimm, 1987; Rotman, 1988; Schleppegrell,

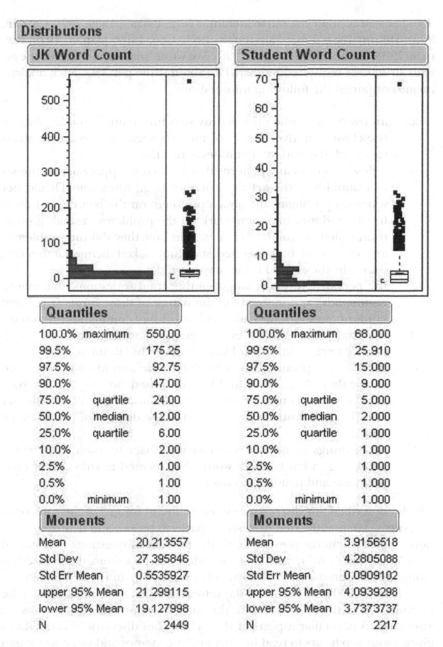

**Figure 2. Quantitative Description of Teacher vs. Student Turn Length.**

2004) as well as other discourse literature (e.g., Barnes, 1969, 1976; Cazden, 2001; Dillon, 1984, 1985; Fairclough, 1995; Gee, 1999; Mehan, 1979; Rotman, 1988; Stubbs, 1975; Wood, 1998). We wrote analytic memos for the activity structures requested by each TR (about three per TR). Each analytic memo contained the following information:

a.  an overview of when the activity structure tended to take place in the classroom, the number of minutes spent on the activity structure each day and total minutes across the weeks;

b.  a description of any patterns that seemed to appear across the set of examples of the activity structure (e.g., when one TR did her warm-up problems, she always put them on the board, read them to the students, had them work on the problems in small groups, then called on volunteers to explain how they did the problem—and when students explained, she always asked them what they did, then why they solved it the way they did);

c.  the pervasive forms of language they used (questions, statements, commands) and more about the nature of those forms (e.g., what kinds of questions they asked, what kinds of commands they used);

d.  the kinds of processes they were engaging students in (e.g., whether they were "thinker" verbs or "doer" verbs [Rotman, 1988]);

e.  the kinds of pronouns they tended to use and when they seemed to use them (e.g., when and how they used "we" vs. "I" and "you" relates to who constructs what kind of knowledge and/or whether knowledge is mutually constructed or the domain of the teacher); and

f.  other things we noticed when we went back to listen to the video tapes (e.g., what kinds of words they seemed to stress, what their wait time and pausing was like).

We also described any discrepant events we found for any of the categories of analysis. We mainly tried to describe patterns we saw and heard, provide percentages whenever possible (e.g., the percent of questions that started with "what" vs. "why"), and then provide examples from the transcripts alongside interpretation to illustrate the discourse patterns we noticed.

In January 2007, we had a full-day retreat at which time we explained the organization of the information in the analytic memos and the reasons for attending to particular aspects of their classroom discourse. Each TR was given about two hours to read his/her analytic memos and was encouraged to write reactions, questions, and alternative interpretations on the memos as he/she read them. We then broke into small groups with one UR and 2–3 TRs in each group, based on the grade levels the TRs taught. In the small groups, the TRs discussed what they thought of the information in the analytic memos, asked clarifying questions of the UR as well as each other, and provided alternative interpretations of and reactions to the data we presented. We spent the final hour of the retreat as a whole group, sharing information from the small group discussions and reactions to the data interpretation. The TRs were asked to start thinking about where, in their data, they saw places that their discourse patterns were either undermining goals they had for their students or places where they thought they could use their discourse patterns more purposefully in order to better achieve their goals. The representation that they used to examine their discourse patterns with respect to their goals was a "beliefs mapping" that we had created during the previous academic year, which I describe in the following section.

## Creating a Beliefs Mapping

Asking the TRs to do a spatial representation of their professed beliefs was guided by work on teachers' beliefs. More specifically, the work of Cooney (Cooney, 2001; Cooney & Shealy, 1997; Wilson & Cooney, 2002) and Chapman (1999, 2002) informed this work. Both articulate spatial representations of "clusters" of beliefs in their work. Drawing on Scheffler (1965), Cooney (2001) contends that it is important to think of beliefs as clusters of "dispositions to act, which include both utterances and actions" (p. 21). The beliefs that teachers draw on (considering the range of beliefs one holds) depend on what is happening at that point in time and with the particular set of students. Cooney and his colleagues argue that the "peripheral beliefs" are the ones that are more amenable to change. In that sense, this conceptualization melds well with doing action research. In fact, Cooney (2001) argues that there are two key elements in changing one's beliefs: "doubt and evidence." The spatial representations allowed the TRs to make their professed beliefs explicit; the action research process allowed them to decide whether their professed beliefs were aligned with their classroom discourse practices.

As part of our project meetings, we often read and discussed short articles or book chapters. After the data collection was completed in March, 2006, we read a chapter from the book *Connecting Mathematical Ideas: Middle School Video Cases to Support Teaching and Learning* (Boaler & Humphreys, 2005). The book includes excerpts from a middle school teacher's (Cathy Humphreys) classroom and each chapter includes a response by university researcher Jo Boaler. In one of the first chapters, Humphreys provides a description of herself and her classroom in order to set the context for the reader. This chapter contains a section in which Humphreys describes "what is closest to the heart" (p. 11) in her teaching. After we read and discussed that chapter, I gave the TRs each a post-it note tablet and asked them, over the next month, to record any words, phrases, or pictures that came to their mind that they felt captured what was closest to their hearts in their teaching. As they did this, the URs analyzed their first two interviews, background information, and three beliefs surveys they completed in the first year of the project. As a result, the URs generated a list of conjectures about what we thought they might write on their post-it notes and recorded them on a set of different colored post-it notes. In April, the URs arrived at the project meeting with large sheets of paper and our conjectures; the TRs arrived with their own set of "professed beliefs" on their post-it notes and their journal. The URs gave each TR a large sheet of paper and asked them to arrange their post-it notes in relationship to the center of the page, which represented what was "closest to their heart" in their teaching.

After the TRs created a spatial representation of their post-it notes, I asked them to write a journal entry explaining what they had written on their post-it notes as well as why they arranged them in the way they had. I then explained that we (the URs) had written conjectures on our own post-it notes and asked them to read each carefully and incorporate our conjectures into their mappings. I told them that they could do one of three things:

1. accept the conjecture and place it in relationship to what they had on their mapping;
2. if they thought the words were close but not quite right, they could change the wording before they incorporated the post-it; and
3. if they disagreed with the conjecture, they could cross it off and place it at the bottom of the large sheet of paper.

The TRs spent about an hour working through our conjectures and then wrote another journal entry that described their reactions to our conjectures and explained what they did with our post-it notes in relationship to their previous mapping. We then had a discussion about the activity and the TRs shared how difficult it was to capture their 'beliefs' on post-it notes.

We gathered the large sheets of paper and the journals the TRs wrote and created smaller electronic versions of them (see Figure 3 for an example) to refer to. We also wrote a set of interview questions for each TR to further probe the words, phrases, and pictures they had written on the post-it notes and to better understand the decisions they made about how to put these in relationship to each other on the paper. Each TR took part in an hour-long interview to explain his/her beliefs mapping and journal entries in more detail. These belief mappings were a crucial tool for reflection for the teachers because they provided a concrete document of the TR's professed beliefs with which they could juxtapose their interpretations of their discourse

prac-
tices,

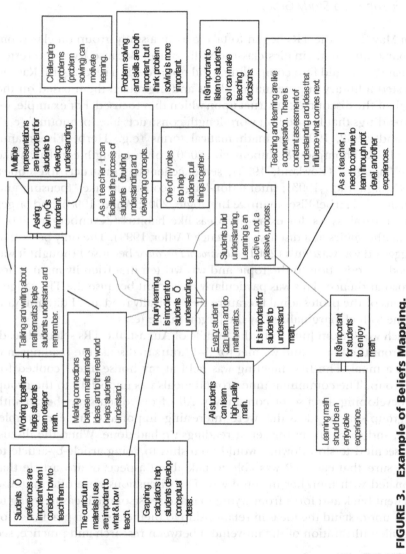

**FIGURE 3.  Example of Beliefs Mapping.**

a point I will explain further in the section on identifying a performance gap.

## Participating in a Study Group

In May 2006, the TRs began to take part in a study group on classroom discourse and mathematics classroom discourse. Because I was interested in what they would find compelling, I did not provide a reading list. Rather, I created a library of potential readings and organized them based on the length of the discourse-construct upon which they focused. For example, we had readings that focused on word-length constructs like "pronouns" (e.g., Rowland, 1992, 1999) or mathematical terms (e.g., Herbel-Eisenmann, 2002; D. R. Thompson & Rubenstein, 2000); phrase-length constructs like some "hedges" (Rowland, 1995); sentence-length constructs like "questions" (e.g., Vacc, 1993); interaction-length constructs like "focusing and funneling" (Herbel-Eisenmann & Breyfogle, 2005; Wood, 1998); and larger socio-cultural issues, for example, ideas like habitus (Zevenbergen, 2001) and mathematics as a discursive practice (Adler, 1999). The only book that I suggested was Cazden's (2001) *Classroom Discourse* because I thought it was a good introduction to the topic and was written in a friendly manner for a broad audience. This was particularly important because the TRs ranged in terms of the professional literature they typically read and I did not want anyone to feel marginalized by the jargon in the texts.[6]

Each week from June until the middle of August, the TRs selected readings from the library and we met for 3–4 hours to discuss the readings and share a meal. The first meeting was held at my house and I cooked for the group. The communal time around meals was important to the group for developing a sense of community. All of our discussions began with the group sharing ideas that were interesting, important, unclear, troublesome, and so on, from the set of readings we had done. When the discussion seemed to slow down, I would then shift to going article-by-article to make sure that each TR was able to talk about aspects of any article that resonated with him/her in some way. The discussions were often intense and went back and forth from trying to make sense of the ideas to trying to better understand the ideas in relationship to their classroom practices (for a detailed illustration of the movement between research and practice, see

---

[6]Needless to say, the TRs found the jargon to be overwhelming and a little amusing at first. For example, they laughed at the fact that people actually named things that we do with language, like "hedging." As they became more familiar with discourse terms, however, they started to use it to comment on their own talk in the project meetings and found it important to their ability to articulate what they were doing with language as well as the differences they were working toward in their classroom discourse.

Herbel-Eisenmann, Drake, & Cirillo, in press). For the set of readings that the teachers found particularly interesting and helpful, see Appendix A.

According to the proposed project plan, the study group on classroom discourse was supposed to end in August, 2006. At that point, however, we had only read some of the articles and book chapters so the TRs requested that we continue with the readings as the school year began. We read literature on classroom discourse until the middle of December and then read two books (Ballenger, 1999; Gallas, 1995) written by elementary teacher-researchers who were working in areas outside of mathematics, but were focusing on their classroom discourse as a segue into a book on doing action research. The main book that guided our action research work was written by Hopkins (2002), but we also read other selections related to action research (e.g., a few chapters from Burnaford, Fischer, & Hobson, 1996; Gallas et al., 1996; Grant & McGraw, 2006). At that point, the TRs expressed concern that most of our readings focused on elementary-level and that few pieces about action research were written by high school mathematics teachers (an exception is Grant & McGraw [2006]). Online searches related to two teacher-research groups that I was aware of (the Boston Teacher Research Group and the Santa Barbara Teacher Research Group) resulted in few options for us to read. After a conversation with Barbara Jaworski, one of my grant advisors, I became aware of a recently-published book that was written about high school classrooms by a woman who had been a high school mathematics teacher, Clare Lee (2006). This book was exactly what the TRs were hoping to read, so our reading shifted back to classroom discourse literature.

## Identifying a "Performance Gap"

In order to move forward, I realized that it was time to start applying some of the ideas in ways that would help the TRs begin to learn to do action research. We collectively decided that the TRs would do a pilot study, allowing them to start exploring the process of action research in a smaller way. Many of them felt a little unclear about what they would be doing and how they would do it. The TRs felt they had learned so much from the readings that it was going to be difficult to get focused. We returned to the Hopkins book and began the process of identifying a "performance gap" or the discrepancies "between behavior and intention" (p. 57). As Hopkins points out, research suggests:

1.  there is often incongruence between a teacher's publicly-declared philosophy or beliefs about education and how he or she behaves in the classroom;

2.  there is often incongruence between the teacher's declared goals and objectives and the way in which the lesson is actually taught; and

3.  there is often a discrepancy between a teacher's perceptions or account of a lesson, and the perceptions or account of other participants (e.g. pupils or observers) in the classroom (p. 57).

Hopkins pointed out that identifying a performance gap can be an important beginning point for cycles of action research.

In order to assist the TRs in identifying their performance gaps, I asked them to bring the discourse analyses we had provided, the readings that they felt were the most helpful to them, their beliefs mapping, and the Hopkins book. We revisited the idea of a "performance gap" and I explained that, initially, I anticipated that their action research projects might grow out of one of three places:

1.  something they were dissatisfied with in their practice prior to coming to the project work;

2.  something they learned from a reading that they had not thought about before; or

3.  a place they could now identify as a mismatch between their beliefs mapping and information we had provided in the discourse analyses or aspects of their practice that they noticed when they watched themselves on the videotapes.

After two hours of revisiting many of the items they brought with them to the meeting, the TRs shared tentative ideas about what they might focus on in their pilot study. Other TRs asked questions, shared ideas, suggested other readings they connected to the ideas, and so on. Each TR recorded his/ her ideas in a journal entry and was asked to keep thinking about this and paying attention to the thing they might focus on over the next week or two.

When the TRs came together for the next project meeting, we asked them to share again what they thought their focus would be and what data they thought they might collect in order to better understand that aspect of their classroom discourse. They also shared their plans for improvement. For example, one TR had written "mathematics is about thinking" on one of her post-it notes and said that interaction patterns she saw both in her analytic memo and in the videos she watched focused more on what answers the students got and whether they were correct or not. She also saw, from the discourse analysis, that many of the processes she referred to in her questions were "doing" verbs. For example, she often asked, "What did you do?" or "How did you do that?" She realized that, if she wanted students to consider mathematics to be a *thinking* process, she had to insert more thinking verbs in her questions and statements. For instance, she thought she could ask her students, "What did you *think about* as you solved that problem?" or reinforce mathematics as thinking by saying things like, "Let's stop and *think* about this problem before we talk about it," or, "Now that we've solved this problem, let's *reflect* on how this solution strategy is related to the last problem we solved."

A shift in clarity seemed to occur when we had to present something at a national conference. Each TR had to select ideas from the readings that s/he felt shaped her/his thinking about the work that we were doing. They visited and revisited articles and selected focus topics to share in the presentation. This also helped them to re-examine their understandings of those readings within the context of their action research projects. The presentations went really well; the audience was engaged and complimentary about the ideas the TRs had. The responses by the audience also seemed to reinforce the fact that the TRs had really important things to say and provided encouragement for them to continue their work.

As we neared the end of the school year, I thought the TRs might not want to continue with the project because it did not feel like we had progressed. Nevertheless, the TRs felt like they finally understood what their focus was and felt better prepared to do action research. They were energized and excited about what they were seeing in their classrooms. In fact, they were excited enough to continue for two more academic years and to each write a book chapter for a book. Herbel-Eisenmann & Cirillo (2009) includes more information about this project, a chapter written by each TR, a set of responses written by some of the authors who inspired and shaped our work, and a concluding chapter that argues for collaborations such as this one to become more common in mathematics education.

## REFLECTIONS FROM A TEACHER EDUCATOR

In this chapter, I described some of the activities I have engaged in with a group of eight teacher-researchers. As I have disseminated some of this work, I sometimes am asked what difference this work makes. Typically, someone poses this question when s/he wants me to connect professional development work to increased student achievement. This is a difficult connection to make and many top scholars are trying to work on methodologies that might allow us to make these kinds of causal links between teacher professional development and student achievement. An underlying assumption of my work is that teachers who think deeply about their practice and who continue to try to make improvements in order to better support their students will have a positive impact on their students' learning. My definition of "positive impact," however, is not reducible to student achievement. In fact, in some other research I have been involved in, we have found that students who maintain their achievement in mathematics can have dramatically different learning dispositions if the learning environment does not fit what they think they need (Jansen, Herbel-Eisenmann, & Smith III, forthcoming).

A colleague once asked me how this work might be transformative, so I asked the TRs how they would respond. They engaged in a lively and interesting discussion about how this work has dramatically changed their thinking and how they now consider many aspects of their practice that were quite invisible to them before their involvement in the project. Furthermore, they now feel that they have a new vocabulary to use to talk about the discourse in their classroom. Sfard (1995) has argued that the reification of processes in algebra has contributed to the development of algebra as a domain. It could be that the naming of discourse moves could play a similar role for teachers as they re-conceptualize their discourse practices. The TRs also recognized that transforming their thinking and naming these language processes was only the beginning. They recognized that transforming their practice was a longer, more difficult process than transforming their thinking. They see this new way of thinking and talking about their practice as the beginning of an ongoing journey.

As Thompson (1992) stated, "We must find ways of helping teachers become aware of the implied rules and beliefs that operate in their classrooms and help them examine their consequences" (p. 142). One way I tried to do this was to work with a group of experienced teachers reflecting on their practice using tools and concepts from discourse analysis, allowing them to examine their enacted beliefs. As a facilitator of this program, however, I sometimes felt like I was not "helping" or "directing" enough. When I would ask the TRs whether the meetings were useful or not, they responded that having the time and space to think was very important. And, being accountable to the group to say something about what they were doing and why was monumental. I think I went into the action research phase of the work thinking that the TRs would have huge epiphanies or that there would be a lot of emotion involved in the sharing. The environment was safe and the TRs shared and celebrated small and big changes. They complimented each other on what they were doing. There were rare instances, however, where I felt like someone's thinking was challenged or really pushed. Most of these instances, in fact, seemed to occur with the least experienced TR in the group, who requested that people push him and reveled in being able to respond. Usually when difficult questions were asked, it was by myself or one of the graduate students. In retrospect, I think I did not do enough to develop the kind of "critical colleagueship" that I had envisioned when I proposed the project. The image that I had of critical colleagueship was based on six elements described by Brian Lord (1994), which included:

1. creating and sustaining productive disequilibrium through self-reflection, collegial dialogue, and on-going critique;
2. embracing fundamental intellectual virtues...openness to new ideas, willingness to reject weak practices or flimsy reasoning... accepting responsibility for acquiring and using relevant information... greater reliance on organized and deliberate investigations... and assuming collective responsibility for creating a professional record of teachers' research and experimentation;
3. increasing the capacity for empathetic understanding...;
4. developing and honing the skills and attributes associated with negotiation, improved communication, and the resolution or competing interests;
5. increasing teachers' comfort with high levels of ambiguity and uncertainty...; and
6. achieving collective generativity (pp. 192–193).

As I reread this list, I can see many of these aspects in the transcripts of the project meetings I have analyzed. There are others, however, that are less apparent. I think the focus on dis-equilibrium and some of the fundamental intellectual virtues as well as increasing teachers' comfort level regarding uncertainty are probably the ones that made me feel as if I was not being as helpful as I would have liked to be. That said, I think I have a different and more complex understanding of the development of such skills and inclinations. I've learned that many of the aspects of critical colleagueship take time, probably more time than the two years that we have worked together doing study group and action research projects. As this particular project nears its end, my own journey continues as I begin work toward more purposeful practices in new collaborations, opening up the activities I have developed through this project to others.

**Acknowledgement:** I would like to thank the teacher-researchers for opening their classrooms to us. I would also like to thank Michelle Cirillo, the graduate research assistant who worked closely with me. We are grateful to the members of the Advisory Board for the project, Tom Cooney, David Pimm, Barbara Jaworski, Cathy O'Connor, and Peg Smith. We appreciate their ongoing support, feedback, and guidance. The discussions with David Wagner, Mary Schleppegrell, Corey Drake, and Jay Lemke provided insights into the professional development work and the analytic methods we employed. This material is based upon work supported by the National Science Foundation under Grant No. 0347906 (Beth Herbel-Eisenmann, PI). Any opinions, findings, and conclusions or recommendations expressed in this material are those of the authors and do not necessarily reflect the views of the National Science Foundation.

## APPENDIX A: READINGS DISCUSSED IN THE PROJECT'S STUDY GROUP

The teacher-researchers were provided with a larger library of discourse-related books, book chapters and articles and selected the following subset of readings to read and discuss (although not in the given order):

Adler, J. (1999). The dilemma of transparency: Seeing and seeing through talk in mathematics classroom. *Journal for Research in Mathematics Education, 30*(1), 47–64.

Arvold, B., Turner, P., & Cooney, T. J. (1996). Analyzing teaching and learning: The art of listening. *The Mathematics Teacher, 89*(4), 326–329.

Ball, D. L. (1993). With an eye on the mathematics horizon: Dilemmas of teaching elementary school mathematics. *The Elementary School Journal, 93*(4), 373–397.

Ballenger, C. (1999). *Teaching other people's children: Literacy and learning in a bilingual classroom.* New York: Teachers College Press.

Bills, L. (2000). Politeness in teacher-student dialogue in mathematics: A socio-linguistic analysis. *For the Learning of Mathematics, 20*(2).

Blunk, M. L. (1998). Teacher talk about how to talk in small groups. In Lampert, M., & Blunk, M. L. (Eds.), *Talking Mathematics in School: Studies of Teaching and Learning.* New York: Cambridge University Press.

Brandell, J. L. (1994). Helping students write paragraph proofs in geometry. *The Mathematics Teacher, 87*(7), 498–502.

Breyfogle, M. L., & Herbel-Eisenmann, B. A. (2004). Focusing on students' mathematical thinking—Not just their responses. *Mathematics Teacher, 97*(4), 244–247.

Cazden, C. (2001). *Classroom discourse: The language of teaching and learning* (2nd ed.) Portsmouth: Heinemann.

Chapin, S. H., O'Connor, M. C., & Anderson, N. C. (2003). *Classroom discussions: Using math talk to help students learn.* Sausalito, CA: Math Solutions Publications.

Cobb, P., Boufi, A., McClain, K., & Whitenack, J. (1997). Reflective discourse and collective reflection. *Journal for Research in Mathematics Education, 28*(3), 258–277.

Dugdale, S., Matthews, J., & Guerrero, S. (2004). The art of posing problems and guiding investigations. *Mathematics Teaching in the Middle School, 10*(3), 140–147.

Forman, E. A., Larreamendy-Joerns, J., Stein, M. K., & Brown, C. A. (1998). "You're going to want to find out which and prove it": Collective argument in a mathematics classroom. *Learning and Instruction, 8*(6), 527–548.

French, P., & MacLure, M. (1983). Teachers' questions, pupils' answers: an investigation of questions and answers in the infant classroom. In Stubbs, M.& Hillier, H. (Eds.), *Readings on language, schools and classrooms* (pp. 193–211). New York: Methuen.

Frykholm, J., & Pittman, M. E. (2001). Fostering student discourse: "Don't ask me! I'm just the teacher!" *Mathematics Teaching in the Middle School, 7*(4), 218–221.

Gallas, K. (1995). *Talking their way into science: Hearing children's questions and theories, responding with curricula.* New York: Teachers College Press.

Gallas, K., Anton-Oldenberg, M., Ballenger, C., Beseler, C., Griffin, S., Pappen-heimer, R., et al. (1996). Talking the talk and walking the walk: Researching oral language in the classroom. *Language Arts, 73,* 608–617.

Glanfield, F., Oviatt, A., & Bazcuk, D. (2006). From teachers' conversations to students' mathematical communications. In Van Zoest, L. R. (Ed.), *Teachers engaged in research: Inquiry into mathematics classrooms, grades 9–12* (pp. 75–95). Greenwich, CT: IAP.

Grant, M., & McGraw, R. H. (2006). Collaborating to investigate and improve classroom mathematics discourse. In Van Zoest, L. R. (Ed.), *Teachers engaged in research: Inquiry into mathematics classrooms, grades 9–12* (pp. 231–251). Greenwich, CT: IAP.

Herbel-Eisenmann, B. A. (2002). Using student contributions and multiple representations to develop mathematical language. *Mathematics Teaching in the Middle School, 8*(2), 100–105.

Herbel-Eisenmann, B. A., & Breyfogle, M. L. (2005). Questioning our *patterns* of questions. *Mathematics Teaching in the Middle School, 10*(9), 484–489.

Herzig, A. H. (2005). Goals for achieving diversity in mathematics classrooms. *Mathematics Teacher, 99*(4), 253–259.

Hopkins, D. (2002). *A teacher's guide to classroom research* (3rd ed.). Berkshire: Open University Press.

Huhn, C., Huhn, K., & Lamb, P. (2006). Lessons teachers can learn about students' mathematical understanding through conversations with them about their thinking. In Van Zoest, L. R. (Ed.), *Teachers engaged in research: Inquiry into mathematics classrooms, grades 9–12* (pp. 97–118). Greenwich, CT: IAP.

Jansen, A. (2006). Seventh graders' motivations for participating in two discussion-oriented mathematics classrooms. *The Elementary School Journal, 106*(5), 409–428.

Kazemi, E., & Stipek, D. (2001). Promoting conceptual thinking in four upper-elementary mathematics classrooms. *The Elementary School Journal, 102*(1), 59–80.

Knuth, E., & Peressini, D. (2001). Unpacking the nature of discourse in mathematics classrooms. *Mathematics Teaching in the Middle School, 6*(5), 320–325.

Lampert, M., Rittenhouse, P., & Crumbaugh, C. (1996). Agree to disagree: Developing sociable mathematical discourse in school. In Olson & Torrance (Eds.), *The handbook of education and human development: New Models of learning, teaching and schooling* (pp. 731–764). Oxford: Basil Blackwell.

Lemke, J. (1990). *Talking science: Language, learning, and values.* Norwood, NJ: Ablex Publishing Corporation.

Manouchehri, A., & Enderson, M. (1999). Promoting mathematical discourse: Learning from classroom examples. *Mathematics Teaching in the Middle School, 4*(4), 216–222.

Manouchehri, A., & Lapp, D. A. (2003). Unveiling student understanding: The role of questioning in instruction. *Mathematics Teacher, 96*(8), 562–566.

Manouchehri, A., & St. John, D. (2006). From classroom discussions to group discourse. *Mathematics Teacher, 99*(8), 544–551.

Martens, M. L. (1999). Productive questions: Tools for supporting constructivist learning. *Science and Children, 24-27,* 53.

Martino, A. M., & Maher, C. A. (1999). Teacher questioning to promote justification and generalization in mathematics: What research practice has taught us. *Journal of Mathematical Behavior, 18*(1), 53–78.

Morgan, C. (1998). *Writing mathematically: The discourse of investigation.* Bristol, PA: Falmer Press.

NCTM. (1991). Discourse standards. In *Professional standards for teaching mathematics* (pp. 34–54). Reston, VA: NCTM.

NCTM. (2000). Communication. In *Principles and standards for school mathematics* (pp. 60–63). Reston, VA: NCTM.

O'Connor, M. C., & Michaels, S. (1996). Shifting participant frameworks: Orchestrating thinking practices in group discussion. In Hicks, D. (Ed.), *Discourse, learning and schooling* (pp. 63–103). New York: Cambridge University Press.

Pimm, D. (1987). *Speaking mathematically.* New York: Routledge and Kegan Paul.

Rittenhouse, P. S. (1998). The teacher's role in mathematical conversation: Stepping in and stepping out. In Lampert, M. & Blunk, M. L. (Eds.), *Talking Mathematics in School: Studies of Teaching and Learning.* New York: Cambridge University Press.

Rowland, T. (1992). Pointing with pronouns. *For the Learning of Mathematics, 12*(2), 44–48.

Rowland, T. (1995). Hedges in mathematics talk: Linguistic pointers to uncertainty. *Educational Studies in Mathematics, 29,* 327–353.

Rowland, T. (1999). Pronouns in mathematical talk: Power, vagueness, and generalization. *For the Learning of Mathematics, 19*(2), 19–26.

Sherin, M. G. (2000). Facilitating meaningful discussion of mathematics. *Mathematics Teaching in the Middle School, 6*(2), 122–125.

Sherin, M. G., Louis, D., & Mendez, E. P. (2000). Students' building on one another's mathematical ideas. *Mathematics Teaching in the Middle School, 6*(3), 186–190.

Sherin, M. G., Mendez, E. P., & Louis, D. (2000). Talking about math talk. In M. J. Burke & F. R. Curcio (Eds.), *Learning mathematics for a new century* (pp. 188–196). Reston, VA: NCTM.

Stein, M. K., & Smith, M. S. (1998). Mathematical tasks as a framework for reflection: From research to practice. *Mathematics Teaching in the Middle School, 3*(4), 268–275.

Thompson, A. G., Philipp, R. A., Thompson, P. W., & Boyd, B. A. (1994). Calculational and conceptual orientations in teaching mathematics. In Aichele, D., & Coxford, A. (Eds.), *Professional development for teachers of mathematics* (pp. 79–92). Reston, VA: National Council of Teachers of Mathematics.

Thompson, D. R., & Rubenstein, R. N. (2000). Learning mathematics vocabulary: Potential pitfalls and instructional strategies. *Mathematics Teacher, 93*(7), 568–574.

Van Zoest, L. R., & Enyart, A. (1998). Discourse, of course: Encouraging genuine mathematical conversations. *Mathematics Teaching in the Middle School, 4*(3), 150–157.

Wagner, D. *If math is a language, how do you swear in it?* Unpublished manuscript.

Whitenack, J., & Yackel, E. (2002). Making mathematical arguments in the primary grades: The importance of explaining and justifying ideas. *Teaching Children Mathematics, 8*(9), 524–527.

Wood, T. (1998). Alternative patterns of communication in mathematics classes: Funneling or focusing? In Steinbring, H., Bussi, M. G. B., & Sierpinska, A. (Eds.), *Language and communication in the mathematics classroom* (pp. 167–178). Reston, VA: NCTM.

Wood, T. (1999). Creating a context for argument in mathematics class. *Journal for Research in Mathematics Education, 30*(2), 171–191.

Wood, T. (2001). Teaching differently: Creating opportunities for learning mathematics. *Theory into Practice, 40*(2), 110–117.

Yackel, E., & Hanna, G. (2003). Reasoning and proof. In Kilpatrick, J., Martin, W. G., & Schifter, D. (Eds.), *A research companion to Principles and Standards for school mathematics* (pp. 227–236). Reston, VA: NCTM.

Zevenbergen, R. (2001). Mathematics, social class, and linguistic capital: An analysis of mathematics classroom interactions. In Atweh, B., Forgasz, H. J., & Nebres, B. (Eds.), *Sociocultural research on mathematics education* (pp. 201–215). Mahwah, NJ: Lawrence Erlbaum Associates.

## REFERENCES

Adler, J. (1999). The dilemma of transparency: Seeing and seeing through talk in mathematics classroom. *Journal for Research in Mathematics Education, 30*(1), 47–64.

Ballenger, C. (1999). *Teaching other people's children: Literacy and learning in a bilingual classroom.* New York: Teachers College Press.

Barnes, D. (1969). Language in the secondary classroom. In *Language, the learner and the school* (pp. 11–77). Baltimore, MD: Penguin Books.

Barnes, D. (1976). *From communication to curriculum.* London: Penguin Books.

Boaler, J., & Humphreys, C. (2005). *Connecting mathematical ideas: Middle school video cases to support teaching and learning.* Portsmouth, NH: Heinemann.

Burnaford, G., Fischer, J., & Hobson, D. (Eds.). (1996). *Teachers doing research: Practical possibilities.* Mahwah, NJ: Lawrence Erlbaum Associates.

Cazden, C. (2001). *Classroom discourse: The language of teaching and learning* (2nd ed.). Portsmouth, NH: Heinemann.

Chapman, O. (1999). Inservice teacher development in mathematical problem solving. *Journal of Mathematics Teacher Education, 2,* 121–142.

Chapman, O. (2002). Belief structure and inservice high school mathematics teacher growth. In Leder, G., Pehkonen, E., & Torner, G. (Eds.), *Beliefs: A hidden variable in mathematics education.* Dordrecht, The Netherlands: Kluwer Academic Publishers.

Cirillo, M., & Herbel-Eisenmann, B. (2006). *Teacher communication behavior in the mathematics classroom.* Paper presented at the North American Chapter of the Psychology of Mathematics Education, Merida, Mexico.

Cooney, T. J. (2001). Considering the paradoxes, perils, and purposes of conceptualizing teacher development. In Lin, F. L., & Cooney, T. J. (Eds.), *Making sense of mathematics teacher education* (pp. 9–31). Dordrecht, the Netherlands: Kluwer Academic Publishers.

Cooney, T. J., & Shealy, B. E. (1997). On understanding the structure of teachers' beliefs and their relationship to change. In Fennema, E., & Nelson, B. S. (Eds.), *Mathematics teachers in transition* (pp. 87–109). Mayway, New Jersey: Lawrence Erlbaum Associates.

Dillon, J. T. (1984). Research on questioning and discussion. *Educational Leadership, 42*(3), 50–56.

Dillon, J. T. (1985). Using questions to foil discussion. *Teaching and Teacher Education, 1*, 109–121.

Doerr, H. M., & Tinto, P. P. (2000). Paradigms for teacher-centered classroom-based research. In Kelly, A. E., & Lesh, R. A. (Eds.), *Handbook of research design in mathematics and science education* (pp. 403–428). Mahwah, NJ: Lawrence Erlbaum Associates.

Fairclough, N. (1995). *Critical discourse analysis.* London: Longman.

Fassnacht, C., & Woods, D. (2005). *Transana v2.0x.* Madison, WI: The Board of Regents of the University of Wisconsin System.

Forman, E. A., Larreamendy-Joerns, J., Stein, M. K., & Brown, C. A. (1998). "You're going to have to find out which and prove it": Collective argumentation in a mathematics classroom. *Learning and Instruction, 8*(6), 527–548.

Gallas, K. (1995). *Talking their way into science.* New York: Teachers College Press.

Gallas, K., Anton-Oldenberg, M., Ballenger, C., Beseler, C., Griffin, S., Pappenheimer, R., et al. (1996). Talking the talk and walking the walk: Researching oral language in the classroom. *Language Arts, 73*(8), 608–617.

Gee, J. P. (1999). *An introduction to discourse analysis: Theory and method.* New York: Routledge.

Grant, M., & McGraw, R. (2006). Collaborating to investigate and improve classroom mathematics discourse. In Van Zoest, L. (Ed.), *Teachers engaged in research: Inquiry into mathematics classrooms, grades 9–12* (pp. 231–251). Greenwich, CT: Information Age Publishing.

Halliday, M. (1978). *Language as social semiotic: The social interpretation of language and meaning.* Baltimore: University Press.

Herbel-Eisenmann, B. (2000). *How discourse structures norms: A tale of two middle school mathematics classrooms.* Michigan State University, East Lansing, MI.

Herbel-Eisenmann, B. (2002). Using student contributions and multiple representations to develop mathematical language. *Mathematics Teaching in the Middle School, 8*(2), 100–105.

Herbel-Eisenmann, B. (2009). Negotiation of the "presence *of* the text": How might teachers' language choice influence the positioning of the textbook? In Remillard, J., Herbel-Eisenmann, B., & Lloyd, G. (Eds.), *Mathematics teachers at work: Inquiries into the relationship between written and enacted curricula* (pp. 134–151). New York: Routledge.

Herbel-Eisenmann, B., & Breyfogle, M. L. (2005). Questioning our *patterns* of questions. *Mathematics Teaching in the Middle School, 10*(9), 484–489.

Herbel-Eisenmann, B., & Cirillo, M. (Eds.). (2009). *Promoting purposeful discourse: Teacher research in mathematics classrooms.* Reston, VA: NCTM.

Herbel-Eisenmann, B., Cirillo, M., & Skowronski, K. (2009). Why classroom discourse deserves our attention! In Flores, A. (Ed.), *Mathematics for every student: Instructional strategies for diverse classrooms, Grades 9–12* (pp. 103–115). Reston, VA: NCTM.

Herbel-Eisenmann, B., Drake, C., & Cirillo, M. (in press). "Muddying the clear waters": Teachers' take-up of the linguistic idea of revoicing. *Teaching and Teacher Education, 25*(2).

Herbel-Eisenmann, B., Lubienski, S. T., & Id-Deen, L. (2006). Reconsidering the study of mathematics instructional practices: The importance of curricular context in understanding local and global teacher change. *Journal of Mathematics Teacher Education, 9*(5), 313–345.

Herbel-Eisenmann, B., Wagner, D., & Cortes, V. (July, 2008). *Encoding authority: Pervasive lexical bundles in mathematics classrooms.* Paper presented at the International Group for the Psychology of Mathematics Education.

Hopkins, D. (2002). *A teacher's guide to classroom research* (3rd ed.). New York: Open University Press & McGraw-Hill.

Jansen, A., Herbel-Eisenmann, B., & Smith III, J. P. (forthcoming). Transitioning out of a reform mathematics program and into high school: Discord and harmony in students' experiences.

Lee, C. (2006). *Language for learning mathematics: Assessment for learning in practice.* New York, NY: Open University Press.

Lemke, J. (1990). *Talking science: Language, learning, and values.* Norwood, NJ: Ablex Publishing Corporation.

Lord, B. (1994). Teachers' professional development: Critical colleagueships and the role of professional communities. In *The future of education perspectives on national standards in America* (pp. 175–204). New York: College entrance examination board.

Mehan, H. (1979). *Learning lessons.* Cambridge, MA: Harvard University Press.

NCTM. (1989). *Curriculum and evaluation standards.* Reston, VA: NCTM.

NCTM. (1991). *Professional standards for teaching mathematics.* Reston, VA: NCTM.

NCTM. (2000). *Principles and standards for school mathematics.* Reston, VA: NCTM.

Pimm, D. (1987). *Speaking mathematically.* London and New York: Routledge and Kegan Paul.

Rotman, B. (1988). Towards a semiotics of mathematics. *Semiotica, 72*(1/2), 1–35.

Rowland, T. (1992). Pointing with pronouns. *For the Learning of Mathematics, 12*(2), 44–48.

Rowland, T. (1995). Hedges in mathematics talk: Linguistic pointers to uncertainty. *Educational Studies in Mathematics, 29,* 327–353.

Rowland, T. (1999). Pronouns in mathematical talk: Power, vagueness, and generalization. *For the Learning of Mathematics, 19*(2), 19–26.

Scheffler, I. (1965). *Conditions of Knowledge.* Chicago: Scott Foresman and Company.

Schleppegrell, M. J. (2004). *The language of schooling: A functional linguistics perspective.* Mahwah, New Jersey: Laurence Earlbaum Associates.

Sfard, A. (1995). The development of algebra: Confronting historical and psychological perspectives. *Journal of Mathematical Behavior, 14,* 15–39.

She, H. C., & Fisher, D. (2000). The development of a questionnaire to describe science teacher communication behavior in Taiwan and Australia. *Science Education, 84*(6), 706–726.

Spillane, J. P., & Zeuli, J. S. (1999). Reform and teaching: Exploring patterns of practice in the context of national and state mathematics reforms. *Educational Evaluation and Policy Analysis, 2*(1), 1–27.

Star, J. R., Herbel-Eisenmann, B. A., & Smith, J. P. (1999). Changing conceptions of algebra: What's really new in new curricula? *Mathematics Teaching in the Middle School, 5*(7), 446–451.

Stein, M. K., Smith, M. S., Henningsen, M., & Silver, E. A. (2000). *Implementing Standards-based mathematics instruction: A casebook for professional development.* New York: Teachers College Press.

Stigler, J. W., & Hiebert, J. (1999). *The teaching gap.* New York: The Free Press.

Stubbs, M. (1975). *Organizing classroom talk* (Occasional paper No. 19). University of Edinburgh: Centre for Research in the Educational Sciences.

Thompson, A. (1992). Teachers' beliefs and conceptions: A synthesis of the research. In Grouws, D. A. (Ed.), *Handbook of research on mathematics teaching and learning.* New York: MacMillan Publishing Company.

Thompson, D. R., & Rubenstein, R. N. (2000). Learning mathematics vocabulary: Potential pitfalls and instructional strategies. *Mathematics Teacher, 93*(7), 568–574.

Vacc, N. N. (1993). Questioning in the mathematics classroom. *Arithmetic Teacher,* 88–91.

Wagner, D., & Herbel-Eisenmann, B. (2008). "Just don't": The suppression and invitation of dialogue in mathematics classrooms. *Educational Studies in Mathematics, 67*(2), 143–157.

Wagner, D., & Herbel-Eisenmann, B. (in press). Re-mythologizing mathematics through attention to classroom positioning. *Educational Studies in Mathematics.*

Wilson, M., & Cooney, T. J. (2002). Mathematics teacher change and development. In Leder, G. C., Pehkonen, E., & Torner, G. (Eds.), *Beliefs: A Hidden Variable in Mathematics Education?* (pp. 127–147). the Netherlands: Kluwer Academic Press.

Wood, T. (1998). Alternative patterns of communication in mathematics classes: Funneling or focusing? In Steinbring, H., Bussi, M. G. B., & Sierpinska, A. (Eds.), *Language and communication in the mathematics classroom* (pp. 167–178). Reston, VA: NCTM.

Zevenbergen, R. (2001). Mathematics, social class, and linguistic capital: An analysis of mathematics classroom interactions. In Atweh, B., Forgasz, H. J. & Nebres, B. (Eds.), *Sociocultural Research on Mathematics Education* (pp. 201–215). Mahwah, NJ: Lawrence Erlbaum Associates.

# CHAPTER 10

# CARE TO COMPARE

## Eliciting Mathematics Discourse in a Professional Development Geometry Course for K–12 Teachers

**Maria G. Fung, David Damcke, Dianne Hart, Lyn Riverstone, and Tevian Dray**

This article describes how the idea of comparing Euclidean geometry with two non-Euclidean geometries (taxicab and spherical) provides participants with engaging mathematical tasks in a professional development geometry course for K–12 teachers. We illustrate with three different examples of how the combination of rich mathematical activities centered on comparison, well-orchestrated group work, and skilled facilitators makes for productive classroom discourse and provides the teacher participants with a model to be emulated in their own classrooms. This model demonstrates for the teacher participants the process of generating and supporting student learning through mathematics discourse in the classroom.

## A FRAMEWORK FOR MATHEMATICS DISCOURSE

The National Council of Teachers of Mathematics (NCTM, 2000) considers communication a fundamental part of doing and of learning mathematics. Rich mathematics discourse is at the center of constructing and connecting knowledge in mathematics. In the November 2007 issue of *The Mathematics Teacher*, which is specifically dedicated to mathematics discourse, Himmel-

*The Role of Mathematics Discourse in Producing Leaders of Discourse*, pages 199–213

berger and Schwartz (2007) and Staples and Colonis (2007) describe three indispensable components behind classroom discourse that promote sense-making and growth in mathematics: an engaging task or activity; collaborative group effort; and a proficient facilitator who knows when to question, when to listen, and when to summarize or clarify ideas. The role of the mathematics instructor is then not only to select challenging and conceptually rich tasks for her class, but also to decide how to organize student work and student reporting in a way that centers on sense-making and on justification of the main mathematical ideas (Groves & Doig, 2004; Krussel, Springer, & Edwards, 2004).

Nathan, Elliam, and Kim (2006) describe how the interplay of convergence and divergence of ideas in the debriefing process generates the type of discourse that helps build students' understanding. Comparison is a natural way of creating this interplay of convergence and divergence of ideas. Rittle-Johnson and Jon R. Star (2007) and Star (2008) argue that comparison is the key component in allowing students to become flexible users of mathematical ideas and techniques and in making connections among concepts. Thus, we believe that grounding instruction in comparison together with the three indispensable components described above can provide new depth to mathematical discourse.

Recent research confirms the expectation that teachers' learning experiences in professional development programs directly impact their own teaching (Heck, Banilower, Weiss, & Rosenberg, 2008). Thus, the design and use of mathematical tasks centered on comparison can also supply teacher participants with an effective model to follow in their own teaching practice as a means of eliciting mathematics discourse.

## THE OREGON MATHEMATICS LEADERSHIP INSTITUTE[1] (OMLI) AND MATHEMATICS DISCOURSE

The Oregon Mathematics Leadership Institute (OMLI) is an NSF-funded Mathematics/Science Partnership that aims at increasing mathematics achievement of K–12 students by providing professional development for in-service teachers from ten participating Oregon school districts. Participants have taken part in three 3-week intensive summer institutes (2005–2007), covering mathematics content as well as leadership skills. Six content courses have been developed, covering Number and Operations, Geometry, Abstract Algebra, Probability and Statistics, Measurement and Change, and Discrete Mathematics; each was offered in 28-hour sessions and designed and delivered by an entire team of mathematics teacher educators.

[1]The Oregon Mathematics Leadership Institute Partnership Project is funded by the National Science Foundation's Math Science Partnership Program (NSF-MSP award #0412553) and through the Oregon Department of Education's MSP program.

One of the main research and pedagogical foci of OMLI has been mathematics discourse, and especially how the quantity and quality of discourse affect student achievement. As a result, the OMLI courses were designed to focus heavily on mathematics discourse, and modeling the use of that discourse, as a method not only for improving the teachers' own conceptual understanding of a variety of mathematical topics, but also as a vehicle for expanding their use of discourse in their own classroom. To accomplish this, all OMLI faculty members were given the opportunity to receive training in discourse facilitation and in collaborative learning as well as in best practices in teaching.

## COMPARING DIFFERENT GEOMETRIES COURSE AT OMLI

The geometry course team consisted of five mathematics educators with diverse backgrounds: a research mathematician, a master teacher, a mathematics educator at an undergraduate institution and two mathematics instructors, one with a Ph.D. in mathematics education, the other with an interest in obtaining that degree.

An interest from the participating OMLI school districts in non-Euclidean geometry led to selecting taxicab and spherical geometries as the two major topics in this course. The two main reasons for choosing these two geometries were that:

1.  both of these topics allowed for hands-on explorations and applications; and
2.  both geometries were unfamiliar to virtually all teacher participants.

Thus the selection of these topics allowed all K–12 participants to begin the course on an equal footing.

One compelling approach to teaching non-Euclidean geometry is to use comparison with Euclidean geometry. This became the fundamental idea behind the course—thus the capstone of the course consisted of creating extensive comparison charts, one for taxicab and Euclidean geometry, and one for spherical and Euclidean geometry. These charts allowed teachers to synthesize their newly acquired knowledge of spherical and taxicab geometries, and to connect it to ideas in the school geometry curriculum.

The *Comparing Different Geometries* course consisted of two separate units—one corresponding to spherical geometry and one corresponding to taxicab geometry, and then a project unit. The learning goals and objectives for the course were for participants to improve their content knowledge of Euclidean geometry, especially as it pertained to undefined terms, axioms, definitions, and justification of geometric results; to focus on effective mathematical reasoning and mathematics discourse; and to make connections between Euclidean and non-Euclidean geometries by examining similarities and differences between them.

In the first unit, spherical geometry, participants investigated the concepts of lines, parallel and perpendicular lines, the set of points equidistant from two given points, common perpendiculars, triangles, squares, and circles.

In the second unit, taxicab geometry, participants explored distance, midpoints, the set of points equidistant from two given points, squares, circles, triangles, and congruence of line segments and triangles.

During the project unit, participants completed independent out-of-class projects. These group projects were investigations of unexplored topics from either spherical or taxicab geometry. Participants made posters of their work and presented the most important parts of their discoveries during the last two days of the course.

Manipulatives were used throughout the geometry course, including colored pencils, rulers, compasses, protractors and grid paper, Etch-a-Sketch toys for demonstrating taxicab distance and Lénárt Spheres with spherical tools for visualizing geometric shapes in spherical geometry.

Each participant was assigned to a group of three or four teachers from different grade levels. Groups were reassigned several times during the course. In conjunction with the groups, a number of norms and protocols were used that were intended to ensure everyone's participation and to encourage risk-taking. These protocols included private time to think, followed by specific instructions about sharing ideas first within a group, and then with the class as a whole (Foreman, 2007).

The three classroom episodes described below were chosen as illustrations of productive mathematics discourse. All of these cases included the generation of new (to the participants) mathematical ideas, making connections among concepts, using precise definitions, and justification of statements. Each of these scenarios is meant to highlight one of the three specific components in the discourse framework described above (engaging task, effective cooperative learning, and skilled facilitation). We believe all three of these episodes (as well as all the remaining ones from the course) were enriched by the comparison nature of the tasks. The comparison technique allowed the teacher participants to see clearly the interconnections among geometric ideas, and the uniquely Euclidean properties of school geometry. Due to the fact that the instructors did not obtain the explicit permission of the participants to use the exact dialogue, the conversations were paraphrased in order to preserve the anonymity of the teachers.

## FUN WITH "SQUIRCLES": FOCUS ON AN ENGAGING MATHEMATICS ACTIVITY

During the unit on taxicab geometry, teacher participants were presented with a task that required them to solve a real-world problem using taxicab geometry. Using an urban context as a starting place led to a rich discussion of the definition of circle, congruent line segments, squares, distance, and area of a circle, allowing participants an opportunity to deepen their understanding of these concepts in both the Euclidean and the taxicab settings, by comparing and contrasting the definitions in the respective geometries. The final part of the activity was for participants to investigate their own questions about circles in taxicab geometry—participants were especially engaged in the task of responding to each other (Figure 1, Appendix A).

The task consisted of the following problems:

a.    Because Clark works early in the morning, he wants to live within three blocks of his workplace. Where could he and Lois live?

b.    Lois does not want to live more than ten blocks from her workplace. For this scenario, where could they live?

c.    How would you describe the boundaries of the regions you drew in problems 1 and 2?

First, the facilitator gave the teacher participants several minutes to think privately about the three questions in the task, after which they shared, in a small-group go-around, their beginning thoughts, ideas, strategies, and/or answers. The small groups then tried to reach a consensus about each of the three questions in the task. The realistic setting of the task captured the attention of the teachers right away, and the unexpected shape of the boundaries they found seemed to intrigue them. A lot of the teacher participants shared that they had no idea that the resulting shape will be built entirely of line segments.

All of the groups were able to agree about the answers to all three parts of the task fairly quickly. Then, the facilitator instructed them to work together to try to write a definition for the set of points that make up the boundary of the regions they found in the task. Because the class had worked to write a precise definition of a circle in Euclidean geometry in a previous unit, most of the groups readily came up with the needed definition as the set of points equidistant from one given point (called the center of the circle).

For a whole-class debrief of this part of the task, the instructor asked a selected participant from one group that had been successful to share their group's definition for a set of points, in taxicab geometry, that looked like the shape shown in Appendix A, Figure 2 in taxicab geometry (the instructor recorded these definitions on a poster).

The first participant offered this definition from the group: The circle with center $O$ and radius $OP$ is the set of all points $Q$, such that $OQ$ is congruent to $OP$. This group had engaged in the task of writing a definition beyond what was asked and begun following their "I wonder..." questions. The teacher–participant shared that their group was wondering whether their circle definition would be satisfied by a square oriented as shown in Figure 3 of Appendix A.

The instructor asked, "What do others think? Is their definition satisfied by this shape?"

Another participant demonstrated that this does not fit the definition of circle because in taxicab geometry $OQ_1$ is not the same length as $OQ_2$ (See Appendix A, Figure 4).

There was still some class confusion about the notion of congruence, in the taxicab sense—this was because the congruent radii of the taxicab circle do not "look" congruent, in the Euclidean sense. The idea that congruence is directly related to the way distances are measured continued to cause disequilibrium in some of the teacher participants for the duration of this activity. But the concrete set-up helped in creating a mental image of the notion of congruence in taxicab geometry.

This discussion was wrapped up when a several participants offered their observations:

1. Euclidean circles are not circles in taxicab geometry; and
2. Not all squares in taxicab geometry are taxicab circles—only the ones whose diagonals line up with the grid.

The class facilitator then asked, "What are some questions that you have about circles? What are you wondering about right now?" Participants took a couple of minutes to write at least two different questions. The instructor noted that, as anticipated, several people had written down the question, "What about $C/d$, the ratio of the circumference to the diameter of a taxi-circle?" and asked groups to start by trying to answer this question. Participants were instructed to try to answer some of the other questions they wrote down only if there was still time after working on the $C/d$ question. The facilitator made this instructional move since calculating this ratio allowed for direct comparison with the Euclidean geometry ratio $\pi$, a topic likely to lead to productive discourse.

To wrap up the discussion of the ratio of circumference to diameter of any taxicab circle, the facilitator first selected two participants to share their results. These participants were chosen based on the quality of their work, the first being concrete and somewhat incomplete while the second was more abstract and precisely justified. The first participant showed one specific example of how to calculate the ratio and then conjectured, without proof, that $C/d = 4$ in general. The second participant gave a general proof by explaining that if the diameter of a taxicab circle is $2r$, then its circumference would be $8r$, and then $8r/2r = 4$. The facilitator at this point proceeded to let the participants decide in their groups on one more question that they would explore.

The discussion that followed in each of the small groups was especially engaging because participants were pursuing their own questions, following whatever path they and their group mates agreed to. After the initial shock that the analogue of Euclidean $\pi$ is in fact equal to 4 in taxicab geometry wore off, most groups chose to examine the formulas for the areas of a circle and a square, again comparing these to the familiar formulas for Euclidean circles and squares.

The class facilitator again selected a participant whose group's question focused on comparison: What is the formula for the area of a taxicab geometry circle, a "squircle," and is it the same as for a Euclidean circle? In response, the group first conjectured that the area of a taxicab circle would be $A = 4r^2$, in complete analogy to the formula for the area of a Euclidean circle. Several different participants disagreed quite vocally and two participants proceeded to offer geometric proofs that the formula for the area of a taxicab circle was in fact: $A = 2r^2$. This was another surprise for the participants and it generated a lot of high-energy discourse not only during class, but also, as we later learned, beyond our classroom, during the subsequent office hour periods.

This classroom vignette presents a concrete illustration of how an interesting applied mathematics task can be combined with effective group protocols and judicious facilitator moves to generate meaningful mathematics discourse centered on precise mathematical definitions and their implications. Simply finding and preparing an interesting mathematics activity was not sufficient for ensuring student learning. The classroom needed to be organized in a way that made each student responsible and responsive, and cooperative learning protocols were one way to achieve this. Every participant was given plenty of time to think by herself, and then there was sharing in the small groups, before sharing with the entire class, the groups were given freedom to pursue their own mathematical questions and the sense of ownership made for high quality work. At each point during the task, everyone was engaged either in doing mathematics, or listening and responding to his classmates.

Furthermore, the class facilitator had to know how to sequence student responses and when to pose a new question. Due to the freedom of participants to explore their own ideas, this activity required a lot of immediate reaction to student ideas and on the fly decisions of which ones to keep and develop further. Moreover, it must be noted that it was the focus (original definition of a taxicab circle) and the re-focus on comparison of circles and their properties (ratio of circumference to diameter, and finding the formula for the area of a taxicab circle) that allowed the participants' discourse to go further and it prompted for justification of conjectures (such as the area of a taxicab circle which was generated in complete analogy to the Euclidean case).

## JIGSAW GROUPS: FOCUS ON EFFECTIVE GROUP WORK

During the final activity on spherical geometry, participants revisited core mathematical ideas from previous class tasks. They completed a chart making explicit the similarities and differences between spherical geometry and Euclidean geometry. The purpose of making these comparisons was to further deepen participants' content knowledge of Euclidean geometry. One way to promote meaningful discourse during this activity was to use a jigsaw protocol. This protocol lent itself nicely to this activity because there were many subtasks to complete, each one corresponding roughly to one geometric object or property to be compared (and thus completing one row in the comparison chart).

The activity began with the facilitator randomly assigning one of the following terms to each small group: lines and parallel lines, perpendicular lines, distance and units, triangles and angle sums, circles and $\pi$, and squares. Appendix B contains the task card that the participants received.

After thinking privately first, group members discussed with each other and came to a consensus about any differences and similarities between Euclidean and spherical geometries for their given term. This information was recorded on the appropriate row of every participant's comparison chart. Participants, then, went through two rounds of rotation to share what their home group came up with, as well as to learn what the other groups did. Each participant took notes on the information he/she gathered. Note that there were 6 topics in the comparison chart (lines and parallel lines, distance, triangles, circles and $\pi$, perpendicular lines, and squares), one per home group. After two rotations, participants returned to their home group and shared what they had learned from the experts in the other groups. They then filled in the remaining rows of the comparison chart.

Due to its structure, this entire activity was discourse driven. It gave each participant a chance to become an expert on a single concept and thus solidify one major geometric topic in her mind. Through discussion in their home groups, participants had to come to a consensus about any similarities and differences they found. Not only did the participants have the opportunity to discuss their ideas, but they also had to listen carefully and then articulate this information to new group members. The use of questioning and justification were emphasized throughout the task.

Furthermore, this mathematical task provided a rich cooperative learning structure. Participants were interdependent and engaged the entire time both in their home and in their subsequent groups by not only becoming experts on one topic but also bringing back new information from the rest of the groups. The comparison theme of this task provided the conceptual thread which made the generated discourse effective.

## SQUARES ON THE SPHERE: FOCUS ON SKILLED FACILITATION

As part of the initial spherical geometry unit, teachers were given the following open-ended task card:

*Investigate Squares on the Sphere. Justify Any Conclusions You Reach During Your Investigations*

The rationale for using this mathematics activity came from the following considerations. Squares are objects familiar to all teacher participants and they provide an example of how the Euclidean definition of a parallelogram with congruent sides and 90° angles needs to be modified to fit with the non-existence of parallel lines on the sphere. The corresponding quadrilaterals on the sphere will have congruent sides and congruent angles, but the angles are necessarily larger than 90°. In the process of attempting to construct squares in spherical geometry, participants focused on the importance of precise and concise definitions in geometry.

This activity was also a suitable choice for group work because it was nonprescriptive and it allowed for a variety of approaches and subtasks to be used in completing it. In our experience, the more open-ended an activity is, the higher the probability of it generating productive discourse during the class discussion.

For this task, teachers sat in groups of three to four members with both early elementary school and late elementary or middle school teachers present in each group. High school teachers were working on a separate investigation at the time, but the entire class participated in the debriefing session.

Participants thought about this task privately for about 10–15 minutes. Then teachers worked with their group on completing the task and making a poster of their results. The posters were displayed on the wall of the classroom or nearby hallway, and groups did a silent walk-through, with one group at a time observing each poster. One person was chosen per group to answer questions about her group's poster, and the rest of the group members read the other posters, got answers about any points that need clarification, and left individual feedback about each poster by writing comments on sticky notes. Then teacher participants went back to their groups and modified their own posters based on the feedback provided by their classmates.

While teachers were busy completing this activity, all the instructors not only observed, asked probing questions (e.g. "Is there another way to think about this type of quadrilateral?" "Do you think this is always true on the sphere?" "Do you have a justification of this statement?") and took detailed notes, but they also reported to the class facilitator for this activity who made a decision about how to organize the debriefing session with the entire class.

The following class began with a careful and deliberate sequencing of the poster presentations based on the facilitator's sense of the flow of ideas that would lead to the most complete understanding of the analogue of squares on the sphere. It was also very helpful that there was enough time for the facilitator to share her ideas with the other instructors of the course and to think some more about her moves. A representative from each selected group was chosen randomly.

The first participant chosen came from a group that got stuck trying to adapt the Euclidean definition of a square as a special parallelogram with congruent sides and 90° angles to the sphere.

The discussion proceeded as follows. (Actual dialogue is paraphrased in order to preserve the anonymity of teacher participants.)

Facilitator: Could you explain how your group approached the square task?

Student 1: Squares have equal sides and 90° angles. But there are no parallel lines on the sphere, so we thought for a long time that there are no squares on the sphere. Then we realized that we were on the wrong track. So the sides of a quadrilateral on the sphere will not be parallel but they might still be equal… and maybe the angles might be equal too.

Facilitator: What did your group do then as a result of this realization?

Student 1: We tried to make a quadrilateral with 4 equal sides and 4 equal angles. We did not know how to do this at first. Suddenly a person in our group realized that perhaps we can start at the North Pole and see if we can find 4 points that are the same distance from this pole. So we wanted the pole to be the center of our spherical square. So we drew a circle centered at the North Pole and by measuring with the spherical compass we found 4 points on the circle that split the circle up into 4 equal parts. Then we connected the 4 points with straight line segments to get our square.

Facilitator: Does anyone have any comments or questions?

Participant: Did you get all your squares in this fashion? We did a similar thing too.

Student 1: Yes, we did.

Facilitator: Now I would like to ask the representative from this (she points) group to come and explain what their group did.

Student 2: We wanted to make a square on the sphere. We started with two perpendicular great circles intersecting at say the North Pole, and then just like the previous group we constructed a series of circles with the North Pole as their center. These circles touched the two great circles in four points that became the vertices of our squares. We constructed a whole bunch of these spherical squares and then measured their angles. We were surprised to find out the angles were not the same—they got bigger, the bigger the square was.

Facilitator: Why did you begin your process with two perpendicular circles?

Student 2: These great circles end up forming the diagonals of the square. I think maybe diagonals of spherical squares have to be perpendicular.

Facilitator: Does anyone see how we can use this?

Participant: Can we define a spherical square as a quadrilateral with congruent perpendicular diagonals?

Facilitator: I encourage everyone to go home and to see if this new proposed definition that a spherical square is a quadrilateral with two equal perpendicular diagonals is equivalent to the definition of a quadrilateral with congruent sides and congruent angles. Now I am inviting the representative of this other group (she points) to come up.

Student 3: We did the same thing as the previous group, but we actually made a chart with the angles of the spherical squares. We made a conjecture that the interior angles are always larger than 90° and smaller than 270°. And you can see here that the larger the area of the square, the larger its angles. This seems to work the same way as for spherical triangles and it is so weird thinking about it together with the usual squares on the plane.

This classroom vignette is just one of many examples of productive mathematics discourse that happened in the course of teaching *Comparing Different Geometries*. We chose to highlight this particular discussion because, due to circumstances that allowed time to think about the sequencing of the participants' presentations, the debriefing part of the discourse went almost flawlessly. The class discussion also illustrated the effectiveness of sequencing in the presentation of mathematical ideas: the class facilitator had a concrete plan of where she wanted to take the discussion and she chose participants to share their ideas in ways that would reinforce the importance of precise mathematical definitions and the extension of mathematical ideas from one paradigm to another.

Even though at first glance it appears that this mathematical task was not centered on comparison, teacher participants began their explorations based on their experiences with Euclidean squares. They attempted to extract the geometric properties that made a square into a square and to apply these properties to the spherical geometry case. Invariably, some of the participants realized that they needed to abandon some of these properties and assumptions in order to fit the situation on the sphere.

Also, the square task was open enough to allow for multiple approaches and ways to succeed, and thus it was truly suitable for independent group investigation. The initial discourse in the small groups or in front of the posters, together with the written poster feedback created opportunities for connecting mathematical ideas and then revisiting them during the revision of the posters. When the final debriefing session was underway, the teacher participants were ready to understand what their classmates did and to clarify in their minds how it related to their own work and to the definition of the spherical square. But it is obvious that comparison underlay this mathematical task and made it possible for the discourse to expand both in magnitude and depth.

## CONCLUSION

We believe that teachers who take mathematics content courses not only improve their knowledge of the discipline but they also tend to absorb the pedagogical tools of their instructors.

The three examples presented in this paper are offered to provide a glimpse of how the OMLI *Comparing Geometry* course was a hands-on discovery-driven course. Mathematics discourse was at the heart of student learning in this course.

We also strongly believe that the comparison method is a very effective tool in teaching non-Euclidean geometry for gains on conceptual understanding. All the participants had some knowledge of Euclidean geometry. Before making comparisons among the concepts of the new geometries, they needed to have a precise understanding of the concepts of Euclidean geometry including the undefined terms, axioms and definitions. Each time a new concept was explored in one of the new geometries, it was important then for them to revisit their knowledge of Euclidean geometry and to have a clear grasp of the same concept in Euclidean geometry, thus, deepening their knowledge of Euclidean geometry while at the same time discovering those same concepts in a new geometry.

But above all, the described classroom examples unraveled in depth mathematics discourse in the experience of all five geometry team members in terms of teacher participants' observations, questioning, generalizations and justification. Throughout the course, participants would make a conjecture based on their knowledge of Euclidean geometry. They would then explore their conjectures and possibly make new conjectures or make generalizations based on their explorations. Finally, they would provide a justification or a proof of their conjecture. This is illustrated particularly in the first activity discussed above. Through exploration, participants made a conjecture that $C/d$ in taxicab geometry is 4. A participant then provided a proof of this fact. In this same activity, the participants first conjectured that the area of a taxicab circle would be $A = 4r^2$ based on the fact that $A = \pi r^2$ in Euclidean geometry. Then through exploration, they conjectured that the area was in fact $A = 2r^2$. It was then noted that two participants offered a geometric justification for the area of a taxicab circle.

These classroom episodes confirmed our belief that the comparison approach allows the discourse in the classroom to deepen, especially when accompanied by rich mathematical tasks, effective participant collaboration, and skilled facilitation. One key to the success of using the comparison approach is that the participants all came with pre-conceived notions about concepts in Euclidean geometry. As they studied other geometries, it was natural for the participants to try to use previous knowledge to investigate new ideas. So the comparison approach was a natural choice to encourage discovery of new ideas since there was a foundation on which to build. Since taxicab and spherical geometries were virtually unfamiliar to all the participants, the opportunities for discourse were greater than if some of the participants had prior knowledge of these geometries. In addition, each participant had varying backgrounds in Euclidean geometry and varying teaching experiences. This led to the opportunity for exciting discourse among the group members. So a foundation on which to build, new geometries, and varying backgrounds of participants all contributed to the success of using the comparison method to encourage discourse.

Throughout the course, the instructors witnessed engaging discourse among the participants. They had three summers in which to make changes in the tasks. After the first summer, it became apparent that open-ended tasks encouraged more discourse and discovery than those tasks which were scaffolded. The geometry instructors also learned to pose questions to the participants when they asked a question rather than immediately answering their questions. This led the participants to have rich discourse among their group members. Thirdly, the geometry instructors learned how to sequence participant presentations in a way that focused on tying together all the different approaches and ideas discovered by the participants. These presentations provided more opportunity for the participants to question and discuss beyond their group discussions. In the end, all of the geometry instructors learned as much about discourse as the teacher participants.

## APPENDIX A

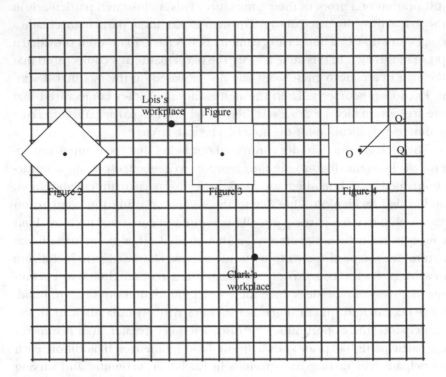

## APPENDIX B: JIGSAW FOR COMPARISON CHARTS

1. In your home group:
2. *Private think time*: Try to write down any similarities and differences between Euclidean and spherical geometries for your group's assigned term on your own–don't discuss with your group mates until everyone is ready.
3. *Go around*: Share your ideas in your home group, all ideas one person at a time.
4. *Discuss*: Come to a consensus about similarities and differences (try to include everyone's ideas.) Make sure everyone in your group understands the ideas discussed. You will be the expert group on your term. Move to your new group, based on the rotation chart.
5. *Go around*: Each person shares what their original group came up with for their assigned term. Listen carefully and ask any clarifying questions you may have. Record the information you learn in the

appropriate row of your comparison chart. Return to your home group.

6. *Go around:* and share what you have learned from each of the experts in the other groups. Be sure to ask any needed clarifying questions. Fill in any of the remaining rows of the comparison chart.

## REFERENCES

Foreman, L. C. (2007). *Collegial leadership and coaching in mathematics toolkit.* West Linn, OR: Teacher Development Group.

Foreman, L. C. (2009). *Comparing different geometries.* Unpublished manuscript, available by request from Teachers Development Group. West Linn. OR.

Groves, S., & Doig, B.(2004). *Progressive discourse in mathematics classes-the task of the teacher.* International Group for the Psychology of Mathematics Education. 28th, Bergen, Norway.

Heck, D., Banilower, E.R., Weiss, I. R., & Rosenberg, S. L. (2008). Studying the effects of professional development: the case of the NSF local systemic change through teacher enhancement initiative. *Journal for Research in Mathematics Education, 39(2),* 113–152.

Himmelberger, S., & Schwartz, D. L. (2007). It's a home run! Using mathematical discourse to support the learning of statistics. *Mathematics Teacher, 101(4),* 257–261.

Krussel, L., Springer, G. T., & Edwards, B. (2004). The teacher's discourse moves: a framework for analyzing discourse in mathematics classrooms. *School Science and Mathematics, 104(7),* 307.

Nathan, M. J., Eilliam, B., & Kim, S. (2006). To disagree, we must also agree: how intersubjectivity structures and perpetuates discourse in a mathematics classroom. *WCER Working Paper No. 2006-6.*

National Council of Teachers of Mathematics. (2000). *Principles and standards for school mathematics.* Reston, VA: Author.

Rittle-Johnson, B., & Star, J.R. (2007). Does comparing solution methods facilitate conceptual and procedural knowledge? An experimental study on learning how to solve equations. *Journal of Educational Psychology, 99(3),* 561–575.

Star, J. R. (2008). It pays to compare! Using comparison to help build students' flexibility in mathematics. *Newsletter of the center for comprehensive school reform and improvement.* April 1.

Staples, M., & Colonis, M.M. (2007). Making the most of mathematical discussions. *Mathematics Teacher, 101(4),* 257–261.

# CHAPTER 11

# SOCIOMATHEMATICAL NORMS IN PROFESSIONAL DEVELOPMENT

## Examining Leaders' Use of Justification and its Implications for Practice

### Rebekah Elliott, Kristin Lesseig, and Elham Kazemi

Leaders of professional development (PD) are asked to engage in ambitious teacher learning if current calls for mathematics reform are going to be advanced in classrooms. Researchers have shown that the effectiveness of mathematics professional development to support teacher learning is directly linked to the quality of facilitation and yet little research has focused on the actual work of PD leaders (Banilower et al., 2006; Even, 2008). In this chapter we report on findings from a leader–development project examining leaders' uses of justification as they engaged in mathematical work and connections between leaders' mathematical work and their work with students and teachers. Our findings show that leaders' uses of mathematical justification built and drew on mathematical knowledge for teaching (Ball et al., 2008). Furthermore, they spoke about how this knowledge was useful for teaching students. However, leaders had difficulty using these ideas to support the teachers with whom they worked. We discuss the implications these findings have for our future research as well as for others designing PD for leaders.

**KEYWORDS:** leaders of professional development, sociomathematical norms, mathematical justification, mathematical knowledge for teaching

*The Role of Mathematics Discourse in Producing Leaders of Discourse*, pages 215–230
Copyright © 2009 by Information Age Publishing
All rights of reproduction in any form reserved.

Classroom practices in which all students are encouraged to make and investigate mathematical conjectures as well as develop and evaluate mathematical arguments, justifications, and proofs are essential for supporting students' development (Ball & Bass, 2003; NCTM, 2000; Yackel & Hanna, 2003). In order to create classroom experiences where all students' mathematical reasoning is central, teachers need to understand what mathematical reasoning and justification entail and how focusing on these may advance students' learning (Knuth, 2002; Lo, Grant & Flowers, 2008; Simon & Blume, 1996; Stylianides & Ball, 2008). Professional development (PD) is one setting in which teachers may come to understand mathematical reasoning and its use for teaching (Ball & Cohen, 1999). To optimize teachers' learning potential in these settings leaders are required to provide opportunities for teachers to explore and justify rich mathematics while simultaneously equipping teachers with the understanding needed to use this mathematical knowledge in classrooms. This mathematically-rich PD requires that leaders both understand mathematics (and certainly a host of other issues on professional learning) and can deploy this understanding in facilitation. A complex task without a doubt. In fact, researchers on PD suggest that the effectiveness of PD to support teacher learning is highly dependent on the quality of facilitation (Banilower, Boyd, Pasley, & Weiss, 2006). Further, researchers have shown that even when mathematics was reported to be the topic of study in PD, many leaders struggled to keep the mathematics content central in activities. Attention was often diverted to other important issues of teaching that did not necessarily support teachers' learning of mathematics (Hill & Ball, 2004; Wilson, 2003; Wilson & Berne, 1999).

Although effective leaders are critical for teacher learning, little attention has been given to what or how PD leaders learn and improve their practice (Ball & Cohen, 1999; Elliott, 2005; Even, Robinson, & Carmeli, 2003). Even's (2008) recent review of the literature on leaders focused instead on the "missing" literature of leader development suggesting what leaders of PD need to know and be able to do in their practice is underdefined and understudied.

Our project, Researching Mathematics Leader Learning (RMLL)[1] attempts to address some of this missing literature by designing and studying learning opportunities for PD leaders. As a major focus of the work in RMLL, this chapter examines leaders' own engagement with mathematical justification. We considered leaders' uses of justification as they engaged in mathematical work and ties between leaders' mathematical work and their work with students and teachers.

[1]RMLL is a five-year National Science Foundation grant (ESI-0554186). Opinions expressed in this chapter are the authors and do not necessarily reflect the views of the granting agent. PI: Judy Mumme; Co-PI :Cathy Carroll; Graduate Research Assistant: Megan Kelley-Petersen.

## RESEARCH QUESTIONS AND CONCEPTUAL FRAMEWORK

RMLL investigates how leaders learn to facilitate mathematically rich PD environments for teaching—environments in which teachers' mathematical learning is central. We have drawn on classroom research to create constructs to support leader learning. One such construct leaders consider as they participate in a series of videocase seminars is sociomathematical norms (Yackel & Cobb, 1996). Sociomathematical norms are the patterns of interaction that guide how mathematics is collectively accomplished in *classrooms*. We have extended this construct to consider what is normative when *teachers* and a *leader* engage in mathematical work in PD. We further discuss our use of this construct below.

This chapter takes up the following research questions for the study of leader learning in RMLL:

1.  What did leaders collectively report as acceptable justification?
2.  What sociomathematical norms for justification were enacted when leaders worked on mathematical tasks?
3.  What ties to practice (teaching and leading) were opened up in leaders' discussions of mathematical justification?

We reasoned that leaders would be guided by what they, themselves, considered to be mathematically convincing, as evidenced in RMLL, when they facilitated PD and pressed for teachers' explanations and justification. Moreover, we conjectured that knowing what leaders pushed for and learned when they were in conversation with one another could provide insight on how leaders worked with teachers' mathematical explanations and justifications to support teacher learning during PD.

### Sociomathematical Norms for Explanation and Justification

Classroom researchers suggest that social and sociomathematical norms guide the nature of the mathematical work that collectively gets accomplished in classrooms (Yackel & Cobb, 1996). For example, social norms might include the expectation that meaningful explanations or justifications are shared and others listen and ask questions in order to make sense of explanations. What counts as an acceptable mathematical explanation or convincing justification for a group are examples of sociomathematical norms (Yackel, 2001). In RMLL seminars, leaders consider the construct of sociomathematical norms to investigate how teachers and a facilitator accomplish mathematical activity in PD.

Researchers have documented the role sociomathematical norms play in pressing all students to think mathematically as well as the impact of norms on student achievement (Kazemi & Stipek, 2001). Little has been done to extend this work into research on teacher learning in professional development. However, it stands to reason that if leaders learn to press teachers and negotiate rich sociomathematical norms, teachers are more likely to learn mathematics.

RMLL has "placed its bets" on the role of studying and developing sociomathematical norms as a rich resource for developing teachers' understanding of mathematics content and for keeping mathematical reasoning central in PD opportunities. As researchers and developers of leaders, we wanted leaders to understand the role productive sociomathematical norms played in PD. These norms, for example, might suggest that explanation or justification focus on mathematical reasoning and uncover mathematical relationships rather than focus on a list of procedures leading to a solution. Our work is explicating what sociomathematical norms might look like in PD.

At the time we designed our seminars, our goals were to help leaders attend to teachers' conceptual over procedural explanations. Our work with sociomathematical norms firmly planted our work in helping leaders develop teachers' mathematical justification. Our intent was to attend to leaders' use of justification so that they could access teachers' justification in our videocase seminars and learn how to negotiate justification in their own PD. However, what we have uncovered in our analysis is that leaders' use of justification resulted in more than accessing others' thinking. Their justifications drew on and developed mathematical knowledge used in teaching (MKT) (Ball, Hill, & Bass, 2005; Hill, Ball, & Schilling, 2008). We note this here to make clear that we did not go into RMLL to work on MKT with leaders. This construct gained traction in the literature only after completing RMLL seminars with leaders. It was through our trying to make sense of leaders' mathematical work and discourse that we uncovered the connection to MKT. This chapter provides readers a view of how we have refined our goals as a result of our analysis of leaders' use of mathematical justification.

## Teacher Learning on Proof and Justification

Over the last decade, research on teachers' understandings of mathematical proof and justification has risen in prominence (Knuth, 2002; Lo, Grant, Flowers, 2008; Simon & Blume, 1996; Stylianides & Ball, 2008). Central in this work is a discussion of what is entailed in proof and justification. Researchers investigating both inservice and preservice teachers' development of classroom proof and justification call for enhanced notions of proof to encompass constructions of mathematical arguments that verify, explain, communicate to others, create new knowledge, and systematize

(Knuth, 2002; Simon & Blume, 1996). One insight we have drawn from this work is that traditional school mathematics' notions of proof have been expanded to include what others might consider mathematical justification. For our analysis, we draw from this research on teachers' work with proof to suggest that justification for leaders has the potential to provide new insights on mathematics, to explain mathematical ideas, and to verify and generalize why something always works.

Our understanding of mathematical justification has been further advanced by the recent work of Lo, Grant, and Flowers (2008) and Grant, Lo, and Flowers (2007), whose research and development work with preservice elementary teachers highlights teachers' justifications as a means to convince and explain. Their work, rooted in research on proof and justification (Simon & Blume, 1996), argues that teachers' justifications need to convince those who do *not* already know mathematics (students) and must explain why a mathematical idea is valid without using compressed mathematical procedures or reasoning. These types of decompressed justifications are very different than typical justifications using procedures or properties that are only convincing if the learner already understands the procedure. They argue that such justifications are important in order for teachers to support students' learning and so that preservice teachers have the opportunities to come to understand why mathematics works.

Grant, Lo, and Flowers (2007) claim that justifications that explicate mathematical reasoning allow teachers to investigate the underpinnings of mathematical arguments and examine parameters on properties and situations. As a result, teachers confront a tendency for justifications to rely on unexamined algorithms or procedures to validate arguments and are pressed to consider why mathematical ideas work. Constructing justifications that explicate reasoning supports teachers' developing mathematical knowledge needed in teaching (MKT) (Ball, Hill, & Bass, 2005; Ball, Thames, & Phelps, 2008). For example, by building justifications that grapple with the mathematical underpinnings of a situation, teachers are building the capacity to examine the accuracy of alternative approaches because they have greater understanding of the mathematical terrain of a task and can follow methods that traverse the landscape of the task. Similar to what Lo and colleagues advance, we have come to understand that a focus in PD on justifications that explicate reasoning may support leaders and teachers learning MKT.

## BACKGROUND ON RMLL

RMLL seminars used tasks and videocases from a larger set of materials developed for PD of mathematics leaders (Carroll & Mumme, 2007). In a series of three, two-day seminars leaders' collective mathematical work on tasks and discussions of their mathematical methods set the stage for the centerpiece of the seminar, a videocase of a mathematics PD leader engaging teachers in the same task. Leaders discussed in small and whole group *what* mathematical explanations were shared in the videocase and *how* participants in the videocase engaged in sharing explanations. These discussions allowed leaders to grapple with PD facilitation issues that were further developed in *Connecting to Practice* activities at the end of each seminar. The trajectory of mathematics tasks in the seminars involved leaders thinking about generalizing as they abstracted from arithmetic, examined geometric patterns, and investigated algebraic reasoning in context.

## PARTICIPANTS' BACKGROUND

Data for this study come from 24 leaders[2] who participated in RMLL seminars over one academic year. RMLL participants were volunteers from a pre-existing *Leader Initiative* (a pseudonym). RMLL seminars took place between years two and three of the *Initiative* when leaders were in the middle of the second academic year of support from the *Initiative* staff. Leaders were forthright in suggesting that they had developed particular ways of doing mathematics prior to attending RMLL seminars.[3] We also concluded that leaders brought with them particular conceptual tools—mathematical knowledge and ways of engaging in mathematical work—that were cultivated over the two summers of the *Leader Initiative*.

All leaders were charged to work with teachers in various mathematics PD contexts—schools, districts, across district, and regionally. However, leaders' job descriptions varied in the degree to which PD responsibilities were a regularly scheduled part of their work. Two-thirds of the leaders worked with teachers across the K–12 spectrum and had one to three years of facilitation experience. One third of leaders had over four years experience as PD leaders. The majority of leaders also worked with K–12 students in some capacity during the day.

## METHODS

Findings were drawn from reviewing data from two videocases, *Convincing Argument* and *Candles*, from the third seminar.[4] Our analysis is focused on

---

[2]RMLL worked with two leader groups. We report here on one group of leaders.

[3]We examine leaders' situated participation so that we can better understand the complexity of issues that are at play in leader learning and what might be needed to support it. We are not making causal claims about the impact of RMLL.

[4]These data are distilled from a large data corpus including: fieldnotes from three video and audiotaped RMLL seminars, artefacts from the seminars, a pre- and post-seminar questionnaire from all leaders, a pre- and post-case study leaders' interview, and fieldnotes and videotape of two PD sessions from case leaders.

seminar three because it was explicitly designed to elicit leaders' thoughts on sociomathematical norms for explanation and justification. Furthermore, it provided a summary of many of the ideas and ways of working on justification that had been observed throughout the seminars. As such, it is illustrative of the seminar series as a whole.

Our analysis progressed in several phases. We reviewed our videodata, corresponding fieldnotes, and leaders' journals from the two videocases with the focus of identifying the sociomathematical norms for justification negotiated by leaders while doing mathematics tasks, examining mathematical solutions, and discussing the videocase. We developed and used an analysis scheme based on our conceptual framework to systematically review the fieldnotes of the third seminar. Initial patterns were identified and then refined by returned to the data looking for confirming and disconfirming evidence. Our findings are based on the patterns identified in the leaders' discussions of mathematical task solutions and the mathematical work captured in the videocases (PD sessions of teachers solving the same tasks).

## FINDINGS

The findings are organized around three claims associated with our research questions. First, in analyzing what leaders collectively reported as acceptable justification and what sociomathematical norms were enacted as leaders worked on mathematical tasks, we claim that leaders' justifications drew on or built their mathematical knowledge for teaching. For our third research question concerning the ties to practice that were made in leaders' discussions of mathematical justification, we found that leaders understood that their mathematical knowledge, developed by constructing justifications, was useful for teaching students. However, leaders reported they struggled to support teachers to take up these same notions of justification and norms for justification. These findings lead us to conclude that PD for leaders needs to:

a. specify how rich norms for justification develop mathematical knowledge for teaching; and
b. explicitly support leaders to take up these ideas in their own PD.

### Leaders' Notions of Justification Entailed Sense-Making of Mathematics

Leaders' explicit talk about explanations made clear they had specific ideas about what counted as justification. In *Convincing Argument* and *Candles*, they said justifications must show mathematical reasoning to make sense of situations. Justification could not just be detailing procedures, "first I did this, then I did this," rather it was to explain and share mathematical relationships so someone could understand the thinking. Leaders' notions of justification were clearly linked to explaining mathematical ideas and communicating to others (Hanna, 1995).

Most leaders reported that generalizations were a part of constructing a justification because they allowed leaders to know that the reasoning "always worked" or was valid beyond specific numbers or for a specific case. Perhaps more important was that generalizations needed to include mathematical reasoning or justification to explain *why* a generalization was valid and provide access for others to understand its validity. One leader offered in the whole group discussion, "to me you definitely have to have the mathematical reason behind your generalization otherwise it's not going to be very productive" (EX). This leader's comment stressed that in order for a generalization to help others understand the mathematical reasoning, generalizations needed justification that explained the underlying mathematics.

To understand the mathematical reasoning underlying a generalization, groups had lengthy conversations stressing the importance of connecting symbolic solutions to other representations (table, graph, visual models, and the situation). One small group acknowledged how a leader's visual model and table were supporting his reasoning on an equation, "…because of your thinking, your process of getting to the formula is why it works as justification because you have every step up to getting to that equation….so by the time you get to the equation, you have justified it…" (ET). Across whole group and small group conversations leaders repeatedly acknowledged that justifications needed to use multiple representations and include a generalization. It was the connections across representations that were powerful for explaining how and why a solution method was mathematically sound.

### Justification Helped Leaders Deepen Their Math Reasoning

Similar to Lo, Grant, and Flower's (2008) notion of justifications that explicate reasoning as a convincing argument, leaders' reported notions of justification supported them to uncover mathematical meaning and validate the mathematical accuracy of their thinking. As a result of this press for justification developed across both the *Leader Initiative*[5] and RMLL, leaders reported they had gained a greater understanding of mathematics. One leader expressed it this way,

---

[5]In our interviews with leaders they suggested that the RMLL focus on mathematics, while not as in-depth as the *Leader Initiative*, supported similar mathematical means and ends. RMLL's mathematical work asked leaders to work mathematics tasks, make sense of others' mathematical thinking, and to think about the mathematical implications for leading PD.

this [math work] has really just pushed me... the way that I might have approached a problem two years ago is totally different from the way I approach
it now in that I try to find my equations in the models...Before I'd just come
up with whatever mechanical way I had been taught to do it. [Now] I look
more for the justification myself (LN).

This leader expressed how symbolic equations had to be tied to other representations (models) to be convinced that she had solved a task. Similarly,
other leaders conferred that their mathematical work on justification had
opened up new ways of thinking about mathematics and as a result they
had a "deeper understanding" of mathematics. We examine this "deeper
understanding" next when discussing leaders' sociomathematical norms
for justification.

### Leaders' Sociomathematical Norms for Justification
### Built and Drew on MKT

To understand what counted as justification for leaders we examined
both small and whole group discussions in two videocases, *Convincing Argument* and *Candles*. *Convincing Argument* focused on what was a sufficient
argument to be convincing for a task that had multiple solution paths, each
potentially providing a generalizable solution.[6] In the *Candles* videocase,
leaders shared their solutions to a series of tasks involving two candles of
varying lengths that burned at different rates (see Appendix A for task).
We found that leaders' notions of justification (what they said) were highly compatible with what constituted justification for the group (what they
did).

#### Leaders' Unpacked the Mathematics

One of the most striking themes that emerged across leaders' mathematical work was a constant push to support any symbolic generalization with
connections to other representations. When leaders solved the *Candles* task
they used tables, physical models, graphs, equations, and words. One leader
commented in small group, "I'm just not willing to accept the fact that this
is something that we have to rely totally on algebra to solve," (NH) as he
pressed his colleagues to makes sense of the mathematics in the *Candles*
task. Leaders wanted to support generalizations with reasoning that was
garnered from connecting to other representations. As a result, they were
making sense of the mathematical relationships in the task to determine
why certain procedures may or may not work in this context.

---

[6]*Convincing Argument* (CA) videocase focuses on a discussion of a task completed in the previous
videocase, Amy's Method (part of RMLL seminar two). CA follows the teachers in the videocase's
discussion of what is a convincing argument or justification for the task.

In *Candles*, leaders' collective focus on the unit rate for each candle allowed them to unpack and uncover the mathematics involved in a related rate task. Given a total time in which each candle would burn but no starting candle height, leaders' reasoned about how to compile a table of values for each candle, how to create a generic visual model of the two candles, and how the situation related to a set of simultaneous equations. The mathematically rich discussion of rates and candle ratios to determine when one candle would be twice the length of the other showed that leaders were working to understand the meaning of rates by examining alternative solutions (as displayed in various representations) and exploring the viability of each solution to advance a justification.

A number of leaders used typical processes for creating an equation[7] to represent the situation, yet this was not left unquestioned. Leaders pressed themselves, and were pressed by others, to examine the meaning underlying the procedures that allowed them to construct the equation. Similarly, leaders determined the solution to the task, but spent most of their time discussing various solution methods using multiple representations and connecting across solutions to determine why the solution made sense. Unlike a student who simply might construct a solution, leaders were focused on unpacking the mathematical meaning in procedures to understand the mathematical terrain of the task.

Our analysis suggested that leaders' justification built or drew on what Ball and colleagues call mathematical knowledge needed for teaching (MKT) (Ball, Hill, & Bass, 2005; Ball, Thames, & Phelps, 2008; Hill, Ball, & Schilling, 2008). In particular, leaders' work with justification drew on or developed specialized mathematical content knowledge (SCK).[8] This is the mathematical knowledge and skills *uniquely* needed by teachers in the conduct of their work. Ball and colleagues identify this as knowledge that allows teachers to interpret and evaluate alternative solutions, analyze mathematics in textbooks, compare affordances of different representations, generate examples, determine if solutions always work, and analyze errors. Leaders' justifications went beyond devising a method to solve the task or settling on a solution. They held one another accountable for developing justifications that used and compared multiple representations, constructed generalizations, and explained mathematical relationships by uncovering mathematical reasoning. Our leaders, unlike Lo et al.'s (2008) preservice teachers who struggled to develop justifications, constructed jus-

---

[7]The typical equation for the situation was $1 - x/9 = 2(1 - x/6)$

[8]Ball and colleagues have identified two kinds of content knowledge that teachers use: Common Content Knowledge (CCK—the mathematical knowledge and skills used by any profession using mathematics) and Specialized Content Knowledge. CCK is the knowledge necessary to correctly do the mathematics in mathematics texts. SCK is the mathematical knowledge entailed by and used in the work of teaching.

tifications that explicated mathematical reasoning and allowed others to be convinced why procedures worked. Moreover, we found that leaders' justifications brought to the fore mathematical knowledge that was utilized in the work of teaching.

## Leaders Struggled to Support Teachers' Understanding and Use of Justification

Unfortunately, leaders also reported that even though they had gained greater understanding of mathematics, valued justifications that explained reasoning, and saw its use for their teaching with students, they struggled to support teachers to take up these notions of justification and generalization in PD. Leaders reported that *their* use of justification had provided insights on teaching students. However, a number of leaders noted that many teachers in PD privileged symbolic algebraic solutions as a means to generalize and were less inclined to connect symbolic solutions to other representations or to acknowledge non-symbolic generalizations as mathematically valuable. Leaders felt less skilled at how to establish similar understandings of justification with teachers in PD and as a result seemed to struggle with making mathematical work in PD seem useful to teachers.

### False Starts Normative for Leaders, but Not Their Staff

Reflecting on leaders' norms for sharing mathematical work in the *Candles* videocase, the RMLL facilitator noted during whole group discussion that leaders shared "false starts" or mathematical strategies that did not advance leaders to a solution. In response, a number of leaders said they had "a lot of practice" sharing struggles with one another in the *Leader Initiative,* similar to their experience in *Candles.* Leaders added that even though "false starts" were the norm for them, they had not been able to cultivate this in their PD with teachers. As a way to make progress on this struggle, the facilitator asked why leaders shared false starts. Leaders responded with comments on their own thinking and how this type of sharing was supportive of them learning to make sense of their own and others' mathematical thinking. The discussion continued with the facilitator specifically linking the moves leaders made that supported the norm of sharing false starts—explaining mathematical reasoning, questioning others' strategies, and connecting across solutions—to providing insights on the groups' sociomathematical norms for explanation.

This topic arose again the second day of the last seminar during the *Connecting to Practice* activity, one meant to directly develop leaders' facilitation practice, when leaders were asked to share issues they faced when teachers shared work. One leader commented that she didn't see a lot of mathematical errors shared in her PD and asked colleagues how they got teachers to publicly share mistakes. At this juncture leaders offered a variety of suggestions and then the conversation returned to why sharing false starts was not the norm with their teachers. We noted, as we traced this conversation across two days, that leaders clearly struggled with how to support teachers sharing thinking that was in process, incomplete, and perhaps included errors. Based on our analysis, we have come to understand these conversations as missed opportunities to make explicit leaders' sociomathematical norms for justification. More importantly, we missed the opening to talk explicitly and systematically about facilitation moves that support these norms and to help leaders develop facilitation strategies that cultivate norms for justification in PD. Finally, we see that we underspecified how we framed sociomathematical norms for justification in PD and its link to leaders' facilitation practice. We return to this idea in our discussion.

### Leaders' Mathematical Practices Didn't Easily Translate to PD Practices

Leaders readily talked about how their deeper understanding of mathematical justification was used to support students' mathematical learning. It was both leaders' talk about how their mathematical knowledge was useful in teaching and the mathematical justifications that we witnessed in RMLL that led us to posit that leaders' justifications built and drew on MKT and in particular specialized content knowledge (SCK). What remained elusive was *how* leaders could take these ideas into their PD practices. For a number of leaders there was a tension between supporting teachers to do math in PD and an undercurrent among their staff that mathematical work was a "waste of time." Leaders saw this tension arise due to the fact that teachers wanted the mathematics they did in PD to have immediate application to the classroom. In hindsight, we now see that we needed to do a better job of helping leaders understand that the mathematical work they did with teachers in PD had direct implications for teachers' work with students. It wasn't something that teachers could "make and take," but involved teachers learning to explicate reasoning so that they could better understand the mathematical work their students were doing.[9] To address the tension leaders described, we needed to link the purpose for doing mathematics to developing teachers' MKT and more specifically to SCK. Although we specified that RMLL leader development work was focused on PD where teachers had the opportunities to develop mathematical knowledge, we underspecified the nature of what that mathematical knowledge looked like and how leaders might attend to it.

---

[9]One way we learned this was by seeing some of our case leaders directly tie the mathematical work in PD to teachers' knowledge needed in teaching. See Kazemi et al. (2009) and Elliott et al. (manuscript) for details.

## IMPLICATIONS AND CONCLUSIONS

A number of questions have arisen for us as we consider leaders' use of justification, our construct of sociomathematical norms, and the implications for PD. In future work, we need to support leaders to consider explicitly: What kinds of justifications might leaders press for when facilitating teachers' mathematical work? What kinds of mathematics tasks might leaders select to support teachers developing MKT through the use of sociomathematical norms for justification? And, how might those tasks look different than the kinds of tasks given to students? We found that leaders needed greater support to consider the role justification played in teacher PD, how it was integral to cultivating productive sociomathematical norms, and the implications of this work for supporting teachers learning MKT.

One means we are exploring to better understand how leaders' work on sociomathematical norms for justification cultivates mathematically rich opportunities for teachers *and* links to developing teachers' MKT is refining the conceptual framing of our work. Recent work by Ball and her colleagues at the University of Michigan has helped us clarify the nature of the mathematical knowledge at play in our efforts (Ball, Thames, & Phelps, 2008; see also: http://sitemaker.umich.edu/lmt/home, http://sitemaker.umich.edu/mod4/home). Initially in our work we specified that the purpose for doing mathematics should be to develop teachers' conceptual knowledge of mathematics. We believe that this initial purpose was too vague to gain traction on doing mathematics and cultivating sociomathematical norms in PD and as a result it hindered our efforts to support leaders. In designing our next cycle of leader development seminars we are considering how the construct of MKT will be highlighted to support leaders' work with teachers.

With our attention on teachers' MKT and SCK in particular as a purpose for doing mathematics, RMLL will make explicit those practices and routines (Kazemi & Hubbard, 2008) that attend to SCK. A few we are considering are: defining reasonable mathematical purposes for PD that are focused on teachers' SCK, constructing and launching mathematics tasks designed to elicit and develop SCK, and pressing for justifications that explicate reasoning. These will be highlighted both in RMLL facilitation moves and in the tools we provide for leaders to clearly support leaders' experiences as learners and facilitators. In addition, leaders may rehearse these practices and routines with colleagues in RMLL.

As a result of our analyses of leaders' work in RMLL, we have come to see that MKT and sociomathematical norms support, and are integral to, establishing productive mathematics-for-teaching discussions. We posit that it is in the negotiation of sociomathematical norms in PD that teachers may have the opportunity to develop SCK and conversely if leaders and teachers have rich understandings of SCK they may press for more robust sociomathematical norms. In our second cycle of work we will explore how the mathematical work leaders do is framed by attending to leaders building their SCK and paying explicit attention to SCK as a purpose for solving, explaining, and justifying mathematics in PD. Our efforts will support the development of leaders' ability to support teacher learning of SCK and advance the paucity of research in the field of leader development.

## APPENDIX A: CANDLES TASK

This task is excerpted from Driscoll (2001):

1.  Maria has a red candle and a green candle. Each candle is 18 cm long. Maria lights each candle at the same time. The red candle takes six hours to burn out, but the green candle takes three.
    - After one hour of burning, which candle is longer? How much longer? Explain how you got your answer.
    - How much time will pass until one of the candles is exactly twice as long as the other? Explain how you got your answer.
2.  Maria lights two candles at the same time. Each candle is 36 cm long. One candle takes three hours to burn out and the other takes six. How much time will pass until the slower-burning candle is exactly twice as long as the faster-burning one? Explain how you got your answer.
3.  Maria lights two candles of equal length at the same time. One candle takes six hours to burn out and the other takes nine. How much time will pass until the slower-burning candle is exactly twice as long as the faster-burning one? Explain how you got your answer.
4.  A blue candle is twice as long as a red candle. The blue candle takes six hours to burn out, and the red candle takes fifteen. After five hours, what fractional part of each candle is left? The blue candle's length is what fraction of the red candle's length? Explain how you got your answer.

## REFERENCES

Ball, D., & Bass, H. (2003). Making mathematics reasonable in school. In Kilpatrick, J., Martin, W.G., & D. Shifter (Eds.), *A research companion to the principles and standards for school mathematics* (pp. 27–44). Reston, VA: National Council of Teachers of Mathematics.

Ball, D. L., & Cohen, D. K. (1999). Developing practice, developing practitioners: Toward a practice-based theory of professional education. In Darling-Hammond, L., & Sykes, G. (Eds.), *Teaching as the learning profession: Handbook of policy and practice* (pp. 3–32). San Francisco: Jossey-Bass.

Ball, D., Hill, H., & Bass, H. (2005, Fall). Knowing mathematics for teaching: Who knows mathematics well enough to teach third grade, and how can we decide *American Educator, 29*, 14–22, 43–46.

Ball, D.L., Thames, M. H., & Phelps, G. (2008). Content knowledge for teaching: What makes it special? *Journal of Teacher Education, 59*(5), 389–407.

Banilower, E. R., Boyd, S. E., Pasley, J. D., & Weiss, I. R. (2006). *Lessons from a decade of mathematics and science reform: A capstone report for the local systemic change through Teacher Enhancement initiative.* Chapel Hill, NC: Horizon Research.

Carroll, C., & Mumme, J. (2007). *Learning to lead mathematics professional development.* Thousand Oaks, CA: Corwin.

Driscoll, M. (2001). *Fostering algebraic thinking toolkit.* Portsmouth, NH: Heinemann.

Elliott, R. L. (2005). *Professional development of professional developers: Using practice-based materials to foster an inquiring stance.* Paper presented at the annual meeting of the American Education Research Association, Montreal, Canada.

Elliott, R. L., Kazemi, E., Lessieg, K., Mumme, J., Carroll, C., & Kelley-Petersen, M. (2009). Conceptualizing the work of leading mathematical tasks in professional development. *Journal of Teacher Education, 60*(4), 364–379.

Even, R. (2008). Facing the challenge of educating educators to work with practicing mathematics teachers. In Wood, T., Jaworski, B., Krainer, K., Sullivan, P., & Tirosh, T. (Eds.), *The international handbook of mathematics teacher education: The mathematics teacher educator as a developing professional* (Vol. 4, pp. 57–74). Rotterdam, The Netherlands: Sense.

Even, R., Robinson, N., & Carmeli, M. (2003). The work of providers of professional development for teachers of mathematics: Two case studies of experienced practitioners. *International Journal of Science and Mathematics Education, 1*, 227–249.

Grant, T. J., Lo, J. J. & Flowers, J. (2007). Shaping prospective teachers' justifications for computation: Challenges and opportunities. *Teaching Children Mathematics, 14*, 112–116.

Hanna, G. (1995). Challenges to the importance of proof. *For the Learning of Mathematics, 15*(3), 42–49.

Hill, H. C., & Ball, D. L. (2004). Learning mathematics for teaching: Results from California's mathematics professional development institutes. *Journal for Research in Mathematics Education, 35*(5), 330–351.

Hill, H., Ball, D., & Schilling, S. G. (2008). Unpacking pedagogical content knowledge: Conceptualizing and measuring teachers' topic-specific knowledge of students. *Journal for Research in Mathematics Education, 39*(4), 372–400.

Kazemi, E., Elliott, R., Lessieg, K., Mumme, J., Carroll, C., & Kelley-Petersen, M. (in press). Doing mathematics in professional development: Working with leaders to cultivate mathematically rich teacher learning environments. In Mewborn, D., & Lee, H. S. (Eds.), *Association of Mathematics Teacher Educators Monograph VI: Scholarly practices and inquiry in the preparation of mathematics teachers.* San Diego, CA: Association of Mathematics Teacher Educators.

Kazemi, E., & Hubbard, A. (2008). New directions for the design and study of professional development: Attending to the coevolution of teachers' participation across contexts. *Journal of Teacher Education, 59*(5), 428–441.

Kazemi, E., & Stipek, D. (2001). Promoting conceptual thinking in four upper-elementary mathematics classrooms. *The Elementary School Journal, 102*(1), 59–80.

Knuth, E. J. (2002). Teachers' Conceptions of Proof in the Context of Secondary School Mathematics. *Journal of Mathematics Teacher Education, 5*(1), 61–88.

Lo, J. J., Grant, T. J., & Flowers, J. (2008). Challenges in deepening prosepctive teachers' understandings of multiplication through justification. *Journal of Mathematics Teacher Education, 11*(1), 5–22.

National Council of Teachers of Mathematics (2000). *Principles and standards for school mathematics.* Reston, VA: Author.

Simon, M. A., & Blume, G. (1996). Justification in the mathematics classroom: A study of prospective elementary teachers. *Journal of Mathematical Behavior, 15,* 3–31.

Stylianides, A. J., & Ball, D. (2008). Understanding and describing mathematical knowledge for teaching: Knowledge about proof for engaging students in the activity of proving. *Journal of Mathematics Teacher Education, 11*(4), 307–332.

Wilson, S. M. (2003). *California dreaming.* New Haven, CT: Yale University Press.

Wilson, S. M., & Berne, J. (1999). Teacher learning and the acquisition of professional knowledge: An examination of research on contemporary professional development. In Iran-Nejad, A., & Pearson, P. D. (Eds.), *Review of Research in Education* (pp. 173–209). Washington D.C.: American Educational Research Association.

Yackel, E. (2001). *Explanation, justification and argumentation in mathematics classrooms.* Paper presented at the Proceedings of the Conference of the International Group for the Psychology of Mathematics Education (25th) Utrecht, The Netherlands.

Yackel, E., & Cobb, P. (1996). Sociomathematical norms, argumentation, and autonomy in mathematics. *Journal for Research in Mathematics Education, 27,* 458–477.

Yackel, E., & Hanna, G. (2003). Reasoning and proof. In Kilpatrick, J., Martin, W.G., & Shifter, D. (Eds.), *A research companion to the principles and standards for school mathematics* (pp. 227–236). Reston, VA: National Council of Teachers of Mathematics.